T0186360

Knowledge and Systems Science

Enabling Systemic Knowledge Synthesis

Knowledge and Systems Science

Enabling Systemic Knowledge Synthesis

Yoshiteru Nakamori

CRC Press
Taylor & Francis Group
Boca Raton London New York

CRC Press is an imprint of the
Taylor & Francis Group, an **informa** business

A CHAPMAN & HALL BOOK

CRC Press
Taylor & Francis Group
6000 Broken Sound Parkway NW, Suite 300
Boca Raton, FL 33487-2742

© 2014 by Taylor & Francis Group, LLC
CRC Press is an imprint of Taylor & Francis Group, an Informa business

No claim to original U.S. Government works

Printed on acid-free paper
Version Date: 20130520

International Standard Book Number-13: 978-1-4665-9300-8 (Hardback)

Library of Congress Cataloging-in-Publication Data

Nakamori, Yoshiteru.
   Knowledge and systems science : enabling systemic knowledge synthesis / Yoshiteru Nakamori.
     pages cm
   Includes bibliographical references and index.
   ISBN 978-1-4665-9300-8 (hardback)
   1. Knowledge, Sociology of. 2. Thought and thinking. 3. Knowledge management. 4. Creative ability. I. Title.

HM651.N35 2013
658.4'038--dc23   2013008341

Visit the Taylor & Francis Web site at
http://www.taylorandfrancis.com

and the CRC Press Web site at
http://www.crcpress.com

Contents

Preface

It is said that new knowledge is created by the interaction between explicit and tacit knowledge. However, this is not a naturally occurring process. New knowledge emerges as a result of the fusion of high-quality information and knowledge in the brain. But this is not necessarily a process that automatically happens. Effort is needed for *systemic knowledge synthesis* to create and justify new knowledge.

How can we synthesize various pieces of knowledge *systemically*? This book tries to answer this question by integrating the ideas in the fields of systems science and knowledge science. Why *systems science*? Because it originated as an academic discipline for solving complex problems by synthesizing various pieces of knowledge. Therefore, in Chapter 1 we will discuss some of the issues of *systemic knowledge synthesis*.

Why *systemic*? Because the whole is not just a collection of parts, but is more than that (Aristotle, 384-322 BC). Analytical methods cannot explain such a phenomenon. In Chapter 2, we will look at *systemic synthesis* by applying two systems methodologies, both of which emphasize the importance of the *human dimension* in problem solving.

However, there is a *paradox* in that systems science, which should be interdisciplinary by its very definition, suffers from a disciplinary split into the soft and hard systems approaches. This situation must be improved if systems science is to define its mission as the synthesis of diversified information, knowledge, opinions, or values.

For this purpose, Chapter 3 will introduce a new integrated systems approach called the *Informed Systems Approach*. As its specific methodology, a knowledge integration model called the *i*-System (knowledge pentagram) will be introduced. This consists of five dimensions for collecting and synthesizing distributed and tacit knowledge.

The *Informed Systems Approach* attempts to integrate soft and hard systems approaches, rather than setting them against each other. It provides for a better understanding between East and West, instead of assuming that they will never meet, and creates new knowledge based on interdisciplinary, or even intercultural integration of knowledge. This idea will be a precursor to the later chapters on knowledge science. Prior to that, in Chapter 4, some mathematical information aggregation techniques will be introduced, which are useful when looking at the scientific dimension of the *i*-System.

A major social change of the early twenty-first century has been called the *knowledge revolution*. Knowledge science has attracted attention as one of the driving forces of this new society. It is hoped that we can dispel the sense of inertia in the integration of knowledge by developing this new discipline. The second half of this book is devoted to the fusion of systems science and knowledge science in order to achieve *systemic integration of knowledge*. In Chapter 5, we will provide an overview of some of the concepts in knowledge science and several approaches to knowledge science.

The organizational knowledge creation theory provides a rational recipe for generating new knowledge, using irrational, or even arational, abilities of the human mind and cultural features of the East. We will discuss, in Chapter 6, several knowledge creation models that have been developed for organizational as well as academic knowledge creation.

Knowledge is constructed and consumed by people in organizations and societies. From this observation, it is clear that no generic model of knowledge is complete without sociological arguments. In seeking a sociological underpinning, Chapter 7 will draw upon the structure-agency-action paradigm, in order to consider a sociological interpretation of the *i*-System.

Historically, there have been two main schools of thinking on knowledge creation. The perspective of grouping is the key distinction between the *context of knowledge discovery* and the *context of knowledge justification*. However, a third approach is required to make knowledge science an academic discipline. With this in mind, in Chapter 8 we will consider how to justify knowledge, and summarize *a theory of knowledge synthesis (construction) systems*.

The science of knowledge science is not *science* in the narrow sense. It is important to acquire a *systemic* view through *trained intuition*, and using methods of justifying new knowledge without simply relying on the scientific method in the narrow sense.

Acknowledgments

First of all I must convey my thanks to those people who gave me the chance to work on such a grand project to create the new discipline of knowledge science. In particular, I have received tremendous support from Professor Shimemura, the former president of the Japan Advanced Institute of Science and Technology, and Professor Nonaka, the first dean of the School of Knowledge Science. Without their direction, I would not have advanced very far in terms of developing knowledge science.

I have been guided by the school regulation: *With the viewpoint of knowledge creation as a practice of nature, individuals, organizations, and society, knowledge science attempts to establish a united discipline of humanities and sciences. The school undertakes education and research to explore the mechanisms of knowledge creation, accumulation and utilization, under an excellent education-research environment. It also nurtures researchers and experienced professionals who have advanced knowledge and application capabilities to lead the knowledge society, as well as broad perspectives and adequate judgment, and advanced communication abilities.*

Next, I would like to thank those who have supported my project from overseas. A little after the school started in 1998, I organized an international symposium on Knowledge and Systems Science, which was held in 2000, and invited many prominent systems scientists. My intention was to establish a new discipline, knowledge science, based on systems science, which is an academic field with a long history. Knowledge science is interdisciplinarity, as is systems science.

Since then we have been organizing this international symposium every year, mostly in Asia, and we started publishing the *International Journal of Knowledge and Systems Science* in 2004. Its current version is now published by IGI Global.

I wish to thank the following scientists, from among many international contributors, for their valuable support: Professor Jifa Cu, the Institute of Systems Science of the Chinese Academy of Sciences, China; Professor Zhongtuo Wang, the School of Management, also the Research Center of Knowledge Science and Technology, at Dalian University of Technology, China; Professor Andrzej Piotr Wierzbicki, the National Institute of Telecommunications, Poland; and Dr. Zhichang Zhu, the University of Hull Business School, UK.

I was responsible for the Center of Excellence program Technology Creation Based on Knowledge Science, in 2003-2008, which was a great opportunity to advance the discipline of knowledge science. Knowledge science is expected to help researchers produce creative theoretical results, not only in management science, but also in important natural sciences such as biotechnology, nanotechnology, the environment, and information technology. For that purpose, it is necessary to design an environment, including time, place, people, context, etc., that supports the development and practice of knowledge creation theory in technology research.

As concrete applications of this program, we started the following educational projects, which have been fed back into the development of knowledge science. The first was a Management of Technology course in which we started teaching knowledge science, especially its application to the management of technology, to local businesspeople in 2004. The second was a lecture course entitled Regional Revitalization Systems, where we also started teaching knowledge science, especially knowledge creation for regional revitalization, to local residents in 2006. The third was an open lecture entitled Innovator Training for Traditional Crafts, in which we started teaching knowledge science in the field of innovation in traditional craft industries in 2008. I would like to thank those people who were involved in the programs or projects described above, who contributed to the development of knowledge science, and who provided me with important knowledge and wisdom.

Finally, I should extend my thanks to the students of the School of Knowledge Science, who listened to my lectures with critical attitudes and helped me develop a course of systems methodology combined with knowledge management. In fact, this book was produced by organizing ten years of my lecture notes from the School of Knowledge Science.

Chapter 1

Issues of Systemic Synthesis

1.1   Yin and Yang

The global environmental issues as a whole are so complicated that it is almost impossible to draw a complete picture (see Figure 1.1). Let us first consider the following three issues:

Issue 1 *Difficulty in observing environments*

Issue 2 *Difficulty in selecting information*

Issue 3 *Difficulty in creating knowledge*

Figure 1.1: Can you complete the puzzle?

Yin and yang are Chinese words meaning dark and light, which are considered always opposite but equal qualities.

1.1.1 Difficulty in observing environments

We know that the global population is rapidly increasing. We also know that resources are rapidly declining. We understand that an increase in carbon dioxide causes global warming. We also understand that we must cooperate to recycle waste. But we wonder whether the global environmental issues are really serious. See Figure 1.2, and read Episode 1 below.

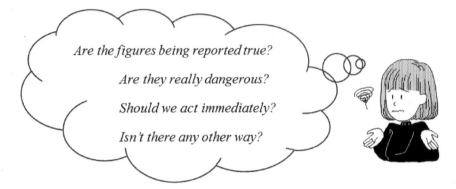

Are the figures being reported true?

Are they really dangerous?

Should we act immediately?

Isn't there any other way?

Figure 1.2: Are global environmental issues really that serious?

Episode 1   Why is the person in this episode angry?

- A telephone call to the municipal office:
    - "I think that the sound of the piano is over the noise limit."
    - "Will you come to measure it?"
- An official comes to investigate the sound:
    - "Well, what can I say?"
    - "It is considerably below the noise limit."
    - "Can you absolutely not stand the sound?"
- The person who complains:
    - "No, the neighbor's wife and her child who plays the piano never greet me when we meet on the road."

How do you interpret this episode?

This person is not only angry about the excessive noise, but also frustrated with the unfriendly attitude of the neighbor's family and the (perceived) unfair situation.

We can understand that there may be two viewpoints in looking at environmental issues. We have to synthesize these two views using *systemic thinking*.

- **Mechanistic view**
    - The meaning of a phenomenon in the research itself cannot be the subject of scientific research.
- **Semantic view**
    - The meaning that we give to the environment varies depending on the context and circumstances.

How can we synthesize these two views while knowing that the following opinions exist?

- If the research results are not tied to our feelings, they may not affect our behaviors.

- Can you detect reality in the prediction of global warming, based on a mathematical model that cannot currently be validated?

- The environment should be understood in terms of the perceptions of reality of the people who live in a particular area.

What kind of world image should we have?

Factual image	$\Longrightarrow$	Hypothetical image
Systematic understanding		*Systemic understanding*

1.1.2   Difficulty in selecting information

We have heard that a lot of data has been and continues to be collected. We have also heard that a lot of models have been created. We can understand somehow that sustainable development is important. We can also understand the necessity of laws and regulations, and international agreements. See Figure 1.3, and read Episode 2 below.

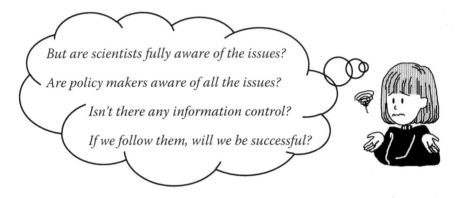

But are scientists fully aware of the issues?

Are policy makers aware of all the issues?

Isn't there any information control?

If we follow them, will we be successful?

Figure 1.3: Can you trust the information on the global environment?

Episode 2   What can we believe?

- A report:

  - If the amount of carbon dioxide in the atmosphere doubles, the temperature will rise by about 2.5 degrees as a result of the greenhouse effect.

- Another report:

  - The Gulf Stream will no longer flow because of melting glaciers in North America, and then the Earth will cool.

- An optimistic guy:

  - Why not? No problem.

How do you interpret this episode?

There is no problem with the global environment.

All seems to be in harmony.

We can understand that there are two stances in evaluating our environment. We have to synthesize these two stances using *systemic thinking*.

- **Reductionist stance**

    - Unless we study in depth and settle individual problems one by one, we cannot solve anything.

- **Holistic stance**

    - When we solve one problem, we face a new problem. Therefore, it is important to achieve balance.

How can we synthesize the two stances, knowing the following difficulties?

- Research methods and contents vary depending on the environmental views of individual researchers.

- To produce excellent research results, the professional environmental scientist inevitably practices reductionism.

- Therefore, the results of individual studies on environmental issues should be synthesized.

- However, the cause of barriers to solving environmental problems is the lack of knowledge and logic to synthesize them.

What should we do other than understanding the problem?

Problem understanding	$\Longrightarrow$	*Problem perception*
Systematic thinking		*Systemic thinking*

1.1.3 Difficulty in creating knowledge

We often hear words such as *recycling* or *zero emissions*. We also hear that a lot of environmentally friendly products are available. We generally understand that we should consider the earth as a whole. We can also understand that we should improve our immediate surroundings as a first step. See Figure 1.4, and read Episode 3.

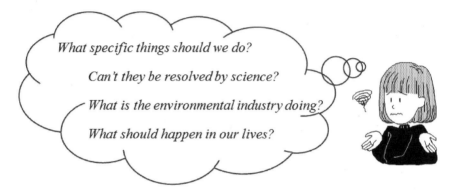

What specific things should we do?

Can't they be resolved by science?

What is the environmental industry doing?

What should happen in our lives?

Figure 1.4: What can we do for global environmental issues?

Episode 3   How should we create new knowledge?

- Young Nakamori (the author of this book):

    – This is an appropriate allocation of instruments to measure the concentration of nitrogen oxides in Kyoto city.

- An official of Kyoto city:

    – The instruments for measuring concentrations of pollutants are to be installed on the roofs of elementary schools.

- Young Nakamori:

    – You should have said that before doing the research.

How do you interpret this episode?

My solution is mathematically perfect.

I do not care about social conditions.

We must understand that we should consider two fields of knowledge. We have to synthesize knowledge from these two fields.

- **Scientific field**

    - By accumulating scientific knowledge, we are creating knowledge of natural science.

- **Social field**

    - By considering social and cultural factors, we are creating knowledge of social science.

How can we synthesize knowledge from two fields, understanding the difficulty that follows in decision making?

- The problem of global warming is an element of a problematique that includes the world economy, energy issues, population issues, etc.

- By understanding complex relationships (observation $\rightarrow$ knowledge) one by one, we can continue our efforts for overall understanding of the world.

- However, even if we can understand the phenomenon, the proper social response (knowledge $\rightarrow$ action) is difficult.

What should we do in addition to knowledge creation?

Knowledge creation		*Culture creation*
Systematic planning		*Systemic planning*

1.2   Discussion of Systemic Thinking

Yin and yang are not opposing forces, but complementary opposites that interact within a greater whole, as part of a dynamic system.

- *Yin and yang* is not a state-oriented categorizing system, but a sense-making metaphor for understanding the infinite possibilities in which new features and patterns of relationships continuously emerge (Cheng, 1997).

- Neither yin nor yang is self-containing: the true yang is the yang that is in the yin; without one, the other is meaningless and seriously lacking. What is significant between opposites is therefore the dynamic, mutual presupposition, transformation, and harmonious interplay in which each opposite complements the other and has its own necessary function; without these joint activities any process would be impossible (Fu, 1997).

Everything has both yin and yang aspects as light cannot exist without darkness and vice versa.

> *We need to synthesize yin and yang with systemic thinking, while paying attention to the fact that they never merge into a synthesis. The loss of opposites means death.*

How can we understand the sentence above, which is somewhat contradictory?

- One of the main theories in *knowledge science* provides a hint: new knowledge is created by the interaction between *tacit knowledge* and *explicit knowledge*. We will consider this theory in later chapters.

- Another hint lies in *systems science*, that is, its fundamental concept of *emergence*, which is the property of a system as a whole that is irreducible to its parts. Let us learn about this concept.

> *A system to us is a set of components, connected such that properties emerging from the system cannot be found in its components.*

With this in mind, answer the following questions.

Question 1.1   Interpret the following sentences by Kant (1724–1804):

- To recognize the truth of things is impossible, because our recognition has to rely on emotional intuition.

- Recognition is not dependent on the object, on the contrary, the object is dependent on the recognition.

- To recognize objects is to *synthesize* them. Nor anything else.

Hint   Human beings do not have the ability to quickly understand complex phenomena analytically. Instead, human beings have the ability to intuitively understand complex phenomena as a whole based on experience-based knowledge. Thus, people recognize the same object differently based on their experience and knowledge. What the object is depends on the recognition of an individual person.

Question 1.2   Interpret the following sentences by Hegel (1770–1831):

- A whole is not in itself a whole.

- A whole will become a whole through its parts.

- However, the parts will never become the whole.

Hint   The whole has the emergent property due to the interactions between the elements, which cannot be explained by the properties of the parts. Because of this, the hierarchical structure is a modeling guide in systems theory; that is, we raise the level of thinking when an emergence occurs.

> *Systemic thinking requires trained insight and intuition to understand the emergent property and the conflicting thoughts.*

1.3   Complexity in Thoughts

Let us consider a regional environmental protection activity to develop a biomass town. A *biomass town* is an area in which a proper profitable use of biomass is or will be executed, by constructing an overall profit use system that links the generation of biomass with its effective utilization with the wide cooperation of all parties concerned.

Kaga city in Ishikawa prefecture, Japan,

constructed a system of promotion which,

however, has not led to an explosive movement.

What are the problems or obstacles?

Kaga city has been recycling food waste as shown in Figure 1.5 (from the homepage of Kaga city).

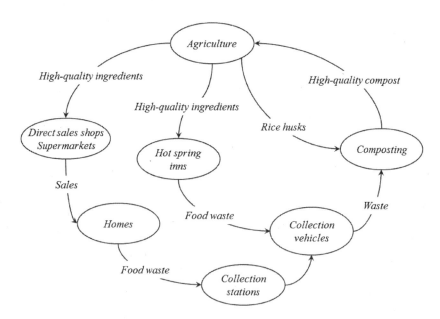

Figure 1.5: A simple model of the food waste recycling process.

The problem can be set out as follows:

Major research question

- What are the issues that interfere with the promotion of the biomass town plan?

Subsidiary research questions

1. What is the current status of the biomass town plan?
2. What do those involved think of the biomass town plan?
3. What do citizens think about the biomass town plan?

Investigation has been performed in accordance with the knowledge integration model[1] as shown in Figure 1.6.

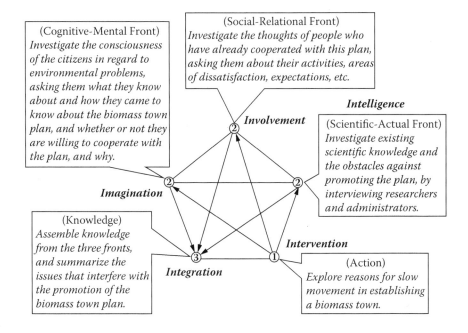

Figure 1.6: A knowledge integration model to synthesize opinions.

---

[1]This model is called the *i*-System (Knowledge Pentagram) and will be introduced later in this book.

1.3.1   Opinions in the scientific-actual front

Interviews with administrative officials and researchers were performed in order to investigate the following:

What is the current status of the biomass town plan?

What is the governmental promotion strategy?

What are the managerial issues in promoting the plan?

What are the technical issues in promoting the plan?

The following are answers from the administrative officials and researchers about the status of the biomass town plan[2] in this city:

Officials

✓ *The city is advertising the plan using various media.*

✓ *But economic benefits are unclear due to the lack of scientific data.*

✓ *There are legal restrictions:*
- *Only selected companies can be involved.*
- *Cooperation between cities is difficult.*

✓ *Possible technical support:*
- *Quality assessment of compost*
- *Production methods for ethanol*

✓ *What we need:*
- *Management of the whole plan*
- *Environmental evaluation*

Researchers

---

[2]The use of biomass is attracting attention around the world for prevention of global warming and for building a recycling-oriented society. Japan has promoted the *Biomass-Japan Strategy* since 2002, inviting local governments to participate in the plan. Kaga was one of the earliest cities to have volunteered, and is now most advanced in food recycling activities.

1.3.2 Opinions in the social-relational front

Interviews with companies and civic organizations were also carried out. Subjects were selected from workers of a resource recycling cooperative, a supermarket council, an agricultural cooperative, a restaurant chain, a cooks' union, a women's council, and a farmers' group.

Are you aware of the biomass town plan?

How do you evaluate the composting business?

What do you think about vegetables grown with food waste?

What are the challenges in promoting the plan?

Opinions are listed as follows; most of them are positive opinions, but there are also some concerns in promoting the plan.

$\boxed{\textbf{Positive opinions}}$

✓ *Desire to support the plan to protect the environment*

✓ *Desire to separate food waste to make high-quality compost*

✓ *Desire to purchase vegetables raised using this compost*

✓ *Belief that those vegetables are quite safe*

$\boxed{\textbf{Matters of concern}}$

✓ *The number of people who will participate in the plan is small.*

✓ *The number of farmers who will use the compost is small.*

✓ *The number of consumers who will buy the vegetables is small.*

✓ *The economic profitability of the plan and the profit return to citizens is unclear.*

✓ *The quality standards and safety standards are still under development.*

✓ *The city should promote the plan more actively.*

✓ *Cooperation between parties is difficult.*

1.3.3   Opinions in the cognitive-mental front

A questionnaire was given to 2,000 citizens[3] living in the city in order to understand the awareness and interest of the citizens:

Are you aware of the biomass town plan?

How can you cooperate with the composting business?

What do you think about the regional recycling system?

Please feel free to tell us your honest opinions about the plan.

The opinions of the citizens were summarized in the affinity diagram,[4] where 165 opinions were eventually reduced to seven groups. These are the seven opinion categories that were viewed as necessary by citizen:

✓ *Future vision of the city to reflect the changes in lifestyle.*

✓ *Significant funding, and a benefit return to residents.*

✓ *Reform the residents' opposition to food waste composting.*

✓ *Good publicity targeting citizens to widely promote the plan.*

Necessities to promote the plan.

✓ *Guidelines to promote this plan from a consumer's viewpoint.*

✓ *Clarification of the significance for residents for participating in this activity.*

✓ *General discussions on the food problem through this initiative.*

---

[3]This survey was carried out in 2008 with a recovery rate of 42%. Answers to the structured questions are omitted here, but most residents wish to contribute to the plan.

[4]The affinity diagram is a business tool used to organize ideas and data, which was devised by Jiro Kawakita (1967) and is often referred to as the KJ Method. See Section 1.5.3.

1.3.4 Main opinions and task diagram

The main points from four groups are summarized as follows:

- Administrators worry about legal restrictions and economic effects, and therefore hesitate to push the plan strongly.

- Scientists have the basic knowledge to produce high-quality compost. However, they are concerned about the overall management of the plan.

- Enterprises want to cooperate in the plan from the standpoint of making a regional contribution, but they are a little hesitant to do business due to the uncertainty of the economic effects.

- Citizens are willing to cooperate with the plan, but they are worried about the future vision, economic, and environmental effects of the plan.

From the above summary, we could draw a task diagram for the four groups as shown in Figure 1.7.

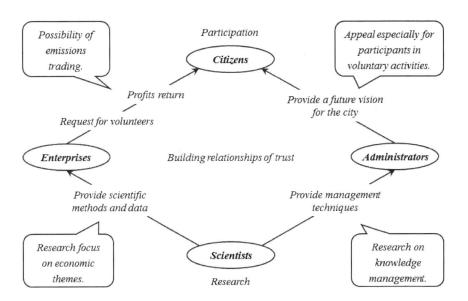

Figure 1.7: A task diagram for promoting the biomass town plan.

1.4   Discussion of Systemic Synthesis

Question 1.3   Synthesize opinions from three fronts in Section 1.3.

Hint   We could identify the problems that are obstacles to promotion of the biomass town plan as follows:

- The lack of scientific basis for economic and environmental effects.

- The lack of business models or management methodologies.

This was actually done by referring to a synthesis table (Table 1.1).

Table 1.1: A table for systemic synthesis of opinions.

Issues ⇕	Scientific-Actual Front	Social-Relational Front	Cognitive-Mental Front	Systemic Synthesis
Economic Benefits	Limited number of companies are engaged.	It is difficult to make a profit from this business.	Methods of returning profits to citizens are unclear.	Lack of management methodology
Environmental Benefits	Scientific data is insufficient for evaluation.	Those involved in the plan wish to improve the environment.	Citizens want a vision of the city that looks to the future.	Lack of scientific information
Technical Problems	Scientists can work together on technical problems.	The quality and safety standards are not established.	Citizens want to ensure the safety of food.	Lack of scientific information
Management	Management methods have not been established.	Fewer people participate in the plan.	The mechanism for participation of citizens is unclear.	Lack of management methodology
Publicity	The city is advertising using some media.	The city should be promoting the plan more actively.	Good publicity is necessary.	Lack of management methodology
Cooperation	Regional cooperation between cities is difficult.	Cooperation between parties is difficult.	Citizens wish to cooperate with the plan.	Lack of management methodology

Synthesis diagram The diagrams in Figure 1.8 support systemic synthesis as intuitive reasoning. In the diagram, SA, SR, and CM stand for *scientific-actual front*, *social-relational front*, and *cognitive-mental front*, respectively.

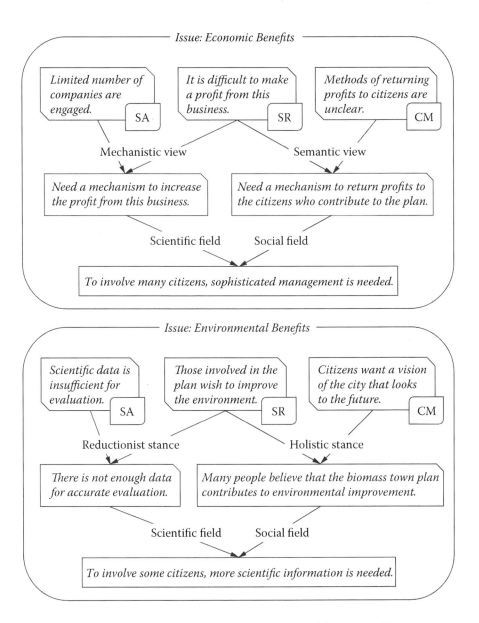

Figure 1.8: Synthesis diagrams to help intuitive reasoning.

Question 1.4 Consider the biomass town plan from three perspectives.

```
┌─────────────────────────────────────────────────────────────┐
│              Mechanistic view vs. Semantic view              │
│                                                              │
│  Factual image of the world  ═══════⟩  Hypothetical image of the world │
└─────────────────────────────────────────────────────────────┘
```

```
┌─────────────────────────────────────────────────────────────┐
│            Reductionist stance vs. Holistic stance           │
│                                                              │
│     Understand problems  ═══════⟩  Perceive problems         │
└─────────────────────────────────────────────────────────────┘
```

```
┌─────────────────────────────────────────────────────────────┐
│   Knowledge in scientific field vs. Knowledge in social field │
│                                                              │
│   Creation of knowledge  ═══════⟩  Creation of culture       │
└─────────────────────────────────────────────────────────────┘
```

Hint

1. Local government has to use taxes properly, and the companies have to pursue profits. So it is understandable that they hesitate to promote things based on a hypothetical image. On the other hand, citizens can think and act through a hypothetical picture. Therefore, this problem is to be promoted through the initiative of citizens, or their representatives, the politicians.

2. Since there are two kinds of complexities intertwined: the complexity of phenomena and the complexity of human minds, it is extremely difficult to solve the problem by following reductionism. The idea that we will not act until we understand things might adversely affect the future. We humans can perceive the problems and act now.

3. As is apparent from the previous considerations, we cannot obtain complete scientific knowledge immediately. Therefore, we are forced to act on the basis of incomplete and uncertain information. In other words, we have to act based on a debate regarding the kind of culture we should build in the future.

1.5 Appendix on Opinion Organizing Methods

Systems science has many tools for idea organization, among them brainstorming, the Delphi method, and affinity diagram will be introduced briefly in this appendix.

1.5.1 Brainstorming

Brainstorming is an idea-generating technique used by a group to find a conclusion for a specific problem by gathering a list of ideas spontaneously contributed by the group members (Osborn, 1953).

Objective:

- Associate something to a certain stimulus.
- Conduct in a group to raise a chain reaction.
- Create a lot of ideas.

Rule:

- First, concentrate on proposing a lot of ideas.
- Perform this in the atmosphere of freedom and fun.
- Do not evaluate ideas that have been put forward.
- Gradually, raise the quality of ideas.

(Team structure)
Leader: Facilitator of the meeting
Secretary: Person to record ideas
Stormers: Persons to generate ideas

(Problems to be treated)
- *Problems that require the judgment "Which is better?" are inappropriate.*
- *Problems that require the idea of "What can I do?" are appropriate.*
- *Problems that are concrete rather than abstract.*

1.5.2   The Delphi method

"The Delphi method is a structured communication technique, originally developed as a systematic, interactive forecasting method that relies on a panel of experts" (Wikipedia, 2013b). It was developed by Project RAND during the 1950-1960s by Helmer, Dalkey, and Rescher. See, for example, Linstone and Turoff (1975), and Rescher (1998).

Objective: Future prediction; for example, technology forecasts by multiple experts.

Procedure:

1. Show the problem to respondents.
2. Respondents fill in answers anonymously.
3. Show the opinion distribution to respondents.
4. Ask them for the answers to the problem again.
5. Repeat the above steps several times.

The purpose of the procedure and the advantages and disadvantages of the Delphi method are shown in Figure 1.9.

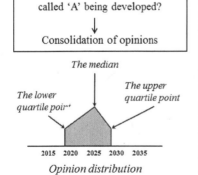

| When do you think of a technique called 'A' being developed? ↓ Consolidation of opinions | *Advantages of the Delphi method:* • It is easy for the respondents to change their opinions. • Their opinions are less affected by influential people. |

The median

The lower quartile point   *The upper quartile point*

2015 2020 2025 2030 2035

Opinion distribution

Disadvantages of the Delphi method:
• The respondents may be irritated by having to answer the problem without enough background.
• The respondents might be influenced by the attached reference materials.

Figure 1.9: Features of the Delphi method.

1.5.3 Affinity diagram

"The affinity diagram is often used in project management and allows large numbers of ideas stemming from brainstorming to be sorted into groups based on their natural relationships, for review and analysis" (Wikipedia, 2013a).

Objective: Develop a framework for the problem (including discovery of sub-problems and definition of a goal).

Procedure: Arrange variables and components, and select evaluation items and evaluation criteria.

1. Clarify the problem.

2. Collect related information (by brainstorming, documentation).

3. Write recommendations for problem solving on cards (one to a card) (small items).

4. Group the cards with similar items (spread the cards on a desk, and decide grouping by looking at all the information).

5. Give names to small groups (medium items).

6. Repeat steps 4 and 5, and make larger groups.

7. Consider the relationships between the groups (similar, conflicting, dependent, causal, or complementary relationships).

8. Make a diagram of groups (enclose each group, and indicate relationships by arrows).

9. Document the structure of the diagram (sub-problems, interactions).

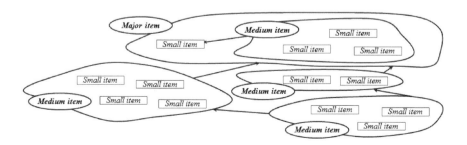

This book emphasizes a dialectic in the sense of confrontational complementarity, instead of a dialectic in the sense of a grand order of thesis, antithesis. and synthesis.

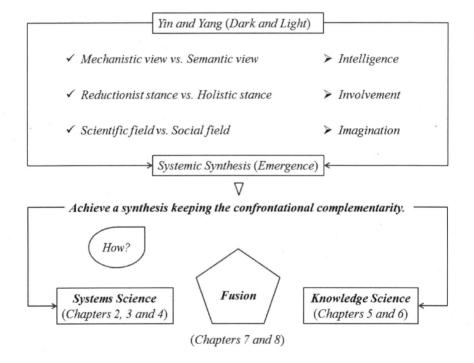

Chapter 2

Systems Approaches

2.1 Complexity and Systems

The *systems approach* is a systematic and organizational approach for considering complex problems based on *systems thinking*, which is recognition of interest focused on the relationships between elements. But, if the number of elements is small or the relationships are simple, the systems approach might not be needed.

> *Systems science is a discipline that deals with complexity.*

See Figure 2.1.

Figure 2.1: How does systems science deal with complexity?

Complexity   Where does the complexity come from? It comes from the perception of humans, as seen in Chapter 1. It also comes from the objects of thought themselves. Let us calculate how rapidly things become complex.

Example 2.1   Things become complex if the number of elements increases. The number of relationships increases in the order of the square of the number of elements, and the number of states increases in the order of the exponent of the number of elements. In fact, if the number of elements is n, the number of binary relationships between two elements is $n(n-1)/2$, and the number of binary states in all elements is 2^n.

Example 2.2   Introduction of symbols:

- Y_t: National economic indicator for the year t

- C_t: Consumption expenditure for the year t

- I_t: Investment expenditure for the year t

- G_t: Government spending for the year t

Consider a simple national economic model that is a simplified version of Flood and Carson (1993):

$$
\begin{aligned}
Y_t &= C_t + I_t + G_t, \\
C_t &= \alpha Y_{t-1}, \\
I_t &= \beta \left(C_t - C_{t-1} \right) + I_{t-1}, \\
G_t &= \gamma Y_{t-1}.
\end{aligned}
$$

If we define the degree of complexity of a model by the number of variables on the right-hand side, which are used to calculate the values of variables on the left, the complexity of this model is 7. But when predicting the variables two years later, five years later, and ten years later, the degree of complexity becomes 13, 75, and 1431, respectively. *Thus, the complexity increases quite rapidly even if the number of direct relationships is rather small.*

Question 2.1   Prove the statements in Example 2.2: When predicting the variables two years later, five years later, and ten years later, the complexity becomes 13, 75, and 1431, respectively. See Section 2.5 for hints.

Emergence and hierarchy The emergence of new properties of a system occurs with an increased level of complexity. Emergent properties are qualitatively different from the properties of the parts of a system. Here, we need *systemic thinking*, and we need to introduce a hierarchy in the system.

> *We raise the level of thinking when an emergence occurs.*

Example 2.3 We can see the academic hierarchy, where a new hierarchy was introduced when an emergence occurred. See Figure 2.2.

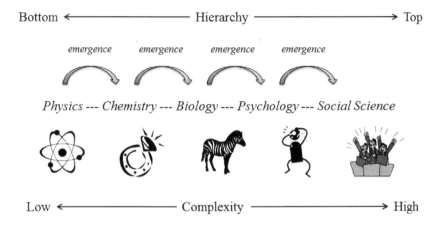

Figure 2.2: Emergence and hierarchy in the disciplines.

- Chemistry was established as an emergence from physics, recognizing that different materials are generated due to differences in the binding of molecules.

- Biology was required because chemistry cannot give a reason why a certain set of molecules starts moving, which is an emergence.

- Psychology was introduced because biology cannot prepare answers to the question of why a human being cries, laughs, or gets angry.

- Social science was developed to explain the behavior of a group of many people, which cannot be explained by the nature of the individuals in the group.

Definition of a system   Through the interaction of its elements, an entire system has specific properties.

> *A system has the emergent property, hierarchy, communication and control processes, and as a whole it can survive against the changing environment in principle (Checkland, 1981).*

Question 2.2   Define the differences between an *education system* and a *mechanism to provide education*. Note that we use *system* in the sense of the above definition, instead of system as an everyday term.

Question 2.3   Interpret the following sentences, which are intrinsic to the definition of a system.

1. The whole is not just a collection of parts but is more than that. (Aristotle, 384-322 BC)

2. Only variety can destroy variety—*the Law of Requisite Variety*[1] (Ashby, 1956, 1958)

Sources   There have been several edited books on systems thinking, which include the latest developments of the discipline at the time of publication (e.g., Buckley, 1965; Emery, 1969, 1981; Beishon and Peters, 1972; Klir, 1991). Midgley (2003) published an edited book that includes a century's worth of important papers from this field.

---

[1]The original meaning is that if a system is to be stable, the number of states of its control mechanism must be greater than or equal to the number of states in the system being controlled. But here, consider the viability or ability of a system against an always changing and complex environment.

2.2 Methodological Complexity

Systems science has been an essential element in the development of industrial civilization and has helped in forming the cultural platform of basic concepts that are essential for knowledge civilization. However, we encounter a basic difficulty, a paradox that *systems science, which should be interdisciplinary by its very definition, suffers from a disciplinary split into the hard systems approaches pursued by the hard sciences and technology and the soft systems approaches pursued by the social sciences.*

Two types of problems A crucial reason for this separation is that in real life situations, we have to deal with two types of problems:

- *Structured problems*:

    - They can be stated explicitly in words. Thus, theories are available for solving them.

    - Example: *How can we carry C from point A to point B at minimum cost?*

- *Unstructured problems*:

    - Even if the existence of the problem is clear in the sense that we feel uneasy, it cannot be explicitly stated other than with a simplistic representation of the situation.

    - Example: *How can we control the flow of information across a company?*

    - Such problems are recognizable but are not defined. Subjective awareness of the problem is always changing with time.

Two schools of thought Therefore, it is no wonder that there are two schools of thought:

- *Hard systems school*:

    - Addresses issues that are clearly defined.

    - Stands on the assumption that systems exist in the perceived world.

- *Soft systems school*:

    – Deals with chaotic, complex, and vague problem situations.
    – Rests on the assumption that the methodology as the process of inquiry can transform itself into a system.

Hard systems school   Computerized mathematical models have been increasingly used for knowledge representation in diverse scientific disciplines, and the hard systems approach becomes understood as being equivalent to the computational science of analyzing such knowledge representations for gaining a deeper understanding of the problems described by such models. The mathematical models used in computational science are intended to be as objective as possible and as exact as needed, but it is obvious that knowledge cannot be absolutely objective and exact.

> *However, hard science and technology cannot succeed without having objectivity and precision as their goals while at the same time understanding their limitations.*

Table 2.1 shows three typical hard systems approaches.

Table 2.1: Hard systems approaches.

Approach	*Origin*	*Purpose*
Systems Engineering	*Bell Telephone Laboratory, in early 1940*	*Design, improvement, and operation of large complicated systems*
Operations Research	*British Troops, in the middle of World War II*	*Operation of systems, based on mathematical analysis*
Systems Analysis	*Rand Corporation, after World War II*	*Design, improvement, and operation of uncertain systems*

Sources   Many books have been published in these fields, as well as new versions of well-known books; for instance, Blanchard and Fabrycky (2010), Taha (2010), and Kendall and Kendall (2010).

Soft systems school   On the other hand, the soft systems school stresses the intersubjective character of knowledge, the role of disputes, the emancipation of all actors taking part in the intersubjective formation of knowledge, and the critical approach characterized by three principles: *critical awareness*, *emancipation and human improvement*, and *pluralism*.

- *Critical awareness* means a critical attitude toward the diverse systems methodologies available.

- *Human improvement* generalizes *emancipation* by insisting that all individuals should be able to best realize their potential.

- *Pluralism* means using diverse systems approaches and methodologies, selected as best suited to a given situation.

Question 2.4

1. What are the main characteristics of complexity in structured situations such as mechanical systems, ecosystems, and economic systems?

2. What are the main characteristics of complexity in unstructured situations such as management, politics, and social issues?

How can we integrate these two approaches? See Figure 2.3.

This book will try to integrate the two approaches, seemingly as different as fire and water. But, before that, let us try to get an overview of movement in systems science; we will see that systems science originally tried to deal with the various fields in a unified manner.

Figure 2.3: How can we integrate the two approaches?

2.2.1 Cybernetics

> Two key contributions to the development of systems science are
> *cybernetics* and *general systems theory.*

The concept of feedback, which is essential for many aspects of contemporary systems thinking, was popularized in Wiener (1948) as an essential part of his concept of *cybernetics.* Incorporating the feedback control by information, *cybernetics* attempted to clarify the mechanism of *homeostasis*[2] in different fields. That is to say, *cybernetics* attempted to:

- use the principle of feedback as the basic concept;

- discover common elements between the mechanical and biological functions; and

- develop a unified theory of control and communication mechanisms.

Wiener studied the control method for identifying flying objects with radar, to predict their course, and to determine how to aim the anti-aircraft artillery. In order to solve this problem, he noticed the importance of the feedback principle based on external information. See Figure 2.4.

The origin of the word *cybernetics* is the Greek word *kybernetes*, which means "the art of steering."

Figure 2.4: The origin of cybernetics.

Sources   *Cybernetics* has attracted the interest of many people even now, more than half a century after it was proposed. See, for instance, Johnston (2008), Pickering (2010), or Medina (2011).

---

[2]Homeostasis is the property of a system that regulates its internal environment and tends to maintain a stable, constant condition of properties (Wikipedia, 2013c).

2.2.2 General systems theory

Bertalanffy (1956, 1968) developed *general systems theory*, stressing the appearance of structural similarities or isomorphism in different disciplines. General systems theory is an interdisciplinary practice that describes systems with interacting components, and is applicable to biology, cybernetics, and other fields. General systems theory also added many new concepts, such as the phenomenon of *synergy* (the whole is bigger than the sum of its parts) and the concept of an open system. Bertalanffy's goals can be summarized as follows:

- Union of sciences

    - It seems impossible that every science is based in physics.

    - Instead, find the same type of law in various fields.

- General systems theory in education

    - Traditional education only trains scientists in individual disciplines.

    - Instead, train general scientists by integrating fields.[3]

- Science and society

    - Physical laws are the technical controls of inanimate nature.

    - What's missing? Knowledge of the laws of human society.

- A person as an individual

    - The true value of humanity:
        → derive value from the mind of the individual;
        → recognize the diversity of knowledge; and
        → assume coexistence with others.

Sources Similar to cybernetics, general systems theory has still attracted a lot of interest in many areas. See, for instance, Skyttner (2001, 2006), Hanson (2002), or Hammond (2003).

---

[3]To train general scientists is not that easy, but it is a goal of the School of Knowledge Science, to be introduced in Chapter 5.

2.2.3   Soft systems thinking

General systems theory acted as a catalyst for soft and social sciences in the critical development of systems science, which resulted in soft systems thinking (for instance, Checkland and Scholes, 1990; Jackson, 2000). Critique from soft systems thinkers was actually directed against operations research,[4] but which resulted in many valuable concepts such as interpretive, emancipatory, and critical systems thinking. See Figure 2.5.

The essential contribution was adding a dimension of human relations, based on the assumption that diverse aspects of human behavior and human relations cannot be adequately modeled mathematically.

Therefore, systems methodologies are needed, which are essentially different sets of systemic procedures and approaches.

Figure 2.5: The essential contribution of soft systems thinking.

Critical systems thinking   *Critical systems thinking* and *critical systems practice* (CST/CSP) present the development of systems methodologies as linear. In order to overcome the perceived failure of hard systems thinking in social problem solving, soft systems thinkers push the frontier of applied systems thinking along two dimensions. The first dimension is *complexity*, and the second is *human relations*.

Dimension of complexity   According to such critiques, original hard systems thinking was able only to deal with simple problems. But both hard systems thinking and the latecomer methodologies along this dimension focus on discovering and handling the objectivity in the system.

---

[4]Classical operations research used mathematical models to describe managerial problems, and prescribed the most efficient solutions to such problems, while the most efficient was predetermined by optimizing, a priori, specified goal functions.

Dimension of human relations   Along the dimension of human relations, soft systems thinkers shift the focus toward intersubjectivity, such as world views with which stakeholders can derive different appreciations of problem situations. According to whether they see differences among both world views and interests as merely stemming from the stakeholders' mental models or as also involving material–structural relations, this cluster of methodologies is further differentiated into soft systems approaches and emancipatory/postmodern methodologies.

Conceptual grid   These two dimensions thus constitute a conceptual grid that allows not only a progressive categorization of methodologies, but also ideal isolation of problem situations, as well as the rational identification of dominant concerns versus secondary problem areas that are to be handled respectively, one-by-one, by a selected, well-defined dominant methodology versus dependent methodologies. During the course of two decades, since the publication of a system of systems methodologies (Jackson and Keys, 1984), the grid has undergone continuous refinements and become more sophisticated (Jackson, 1991, 2000, 2003). See Figure 2.6.

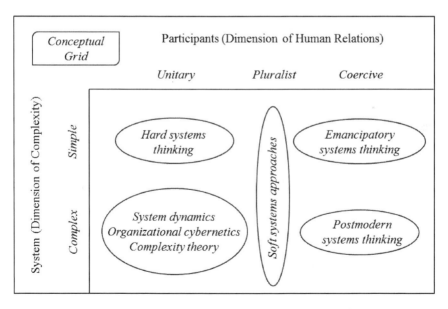

Figure 2.6: A system of systems methodologies (SOSM).

Technical terms   Some of the technical terms in Figure 2.6 are explained below.[5]

- Dimension of human relations

    - *Unitary*: Participants have similar values, beliefs, and interests.

    - *Pluralist*: Although their basic interests are compatible, participants do not share the same values and beliefs.

    - *Coercive*: Participants have few interests in common and would hold conflicting values and beliefs. Decisions are taken on the basis of who has the most power.

- Emancipatory systems thinking

    - *Emancipation*: Are disadvantaged groups being assisted in getting the things to which they are entitled?

    - Example: Critical systems heuristics (Ulrich, 1983)

    - Example: Team syntegrity (Beer, 1994)

- Postmodern systems thinking

    - Seek to promote diversity in problem resolution. Justify interventions on the basis of exception and emotion.

    - Example: Participatory appraisal of needs and the development of action (Taket and White, 2000)

Question 2.5

1. Consider an example of feedback control.

2. What are the aims of cybernetics and general systems theory, respectively?

3. How did the founders of soft systems approaches criticize the so-called hard systems approaches?

4. What did they newly add to the systems approach?

---

[5]Some systems approaches such as *organizational cybernetics* and *critical systems heuristics* will be explained in the appendix of this chapter.

2.3   Soft Systems Methodology

The aim of applying *soft systems methodology* (SSM) (Checkland, 1981) is, to some extent, the reform of an unsatisfactory situation. Such reforms should be *systematically desirable* and *culturally feasible*. The SSM is not necessarily intended to achieve a consensus of the people involved, but intended to activate discussion for a compromise. The process of this methodology is not a systematic procedure to give a correct objective answer; rather it is an emergent, *systemic process* so that the user can continue exploratory learning. Figure 2.7 shows the process of SSM.

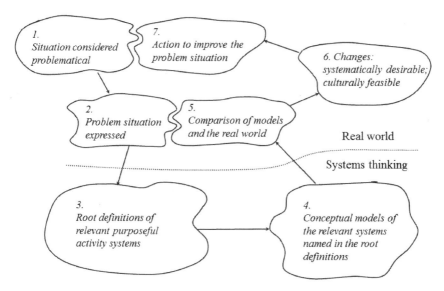

Figure 2.7: The process of soft systems methodology (Checkland, 1981).

2.3.1   Process of soft systems methodology

Stage 1   (*Situation considered problematical*) The manager of a hamburger shop thought that the number of customers had been decreasing probably because of the recent bad economy. He then decided to make an activation plan using SSM.[6]

---

[6]This example was given by Kyoichi Kijima, Tokyo Institute of Technology.

Stage 2 (*Problem situation expressed*) In order to clarify how the people involved think about the store, the manager should ask them what kind of system the store represents for them. He should receive subjective opinions of the store in response. He would receive a variety of answers from different perspectives. What is important here is that the answers are all views related to the store, so it does not make sense to consider whether the opinions are good or bad.

- From customers, it is a system that provides
    - a place to kill time;
    - a place to meet a boyfriend;
    - a place for young mothers to talk to each other, while giving their kids something to eat;
    - a place to read books; or
    - a place for workers to have a short rest before going home, and to reflect on what happened at work that day.

- From employees, it is a system that provides
    - a carefree place for a part-time job; or
    - a place with the opportunity to become friends with girls.

- From the store manager, it is a system that welcomes many customers into the store and increases sales.

Stage 3 (*Root definitions of relevant purposeful activity systems*) In response to the problems recognized in Stage 2, in Stage 3, *root definitions* are formulated to improve the problem situation. Root definitions are types of worldviews and philosophies that reflect the perceptions of various people involved in the problem situation. The same number of root definitions are created as the number of problems recognized in Stage 2. For instance, the burger shop should be a place for workers to have a short rest before going home, and to reflect on what happened at work that day. This is related to values, or the belief that the burger shop serves not only snacks but is also a place where people can relax. Under its values, the burger shop will be represented, for example, as an input–output system as shown in Figure 2.8.

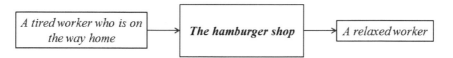

Figure 2.8: The hamburger shop can be an ideal place for workers.

In this type of worldview, not only the taste of the burger, but the atmosphere of the shop has important implications.

Stage 4 (*Conceptual models of the relevant systems*) In Stage 4, *a conceptual model*, expressed as a diagram, is constructed from the root definition. While in the root definition a human activity system is represented as a black box, a conceptual model is a white box represented by a chain of conversion processes. See Figure 2.9.

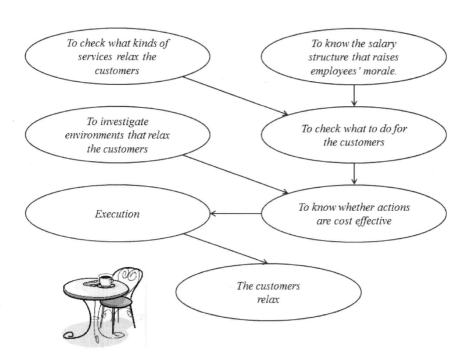

Figure 2.9: An example of the conceptual model.

Stage 5   (*Comparison of models and the real world*) In Stage 5, these models are compared with the perceived actual problem situations considered in Stage 2. The aim is to provide material for debate about possible change among those interested in the problem situation.

Stage 6   (*Changes: systematically desirable and culturally feasible*) Stage 6 should see the concerned actors think about what changes, if any, are both desirable in terms of the models and feasible given the prevailing history, culture, and politics. Changes should be *systematically desirable* for the whole system. However, this alone is not enough. What is more important is to find *culturally feasible* changes in a particular situation.

Stage 7   (*Action to improve the problem situation*) When some solutions are found, action can be taken that alleviates some of the initial unease, and therefore improves the problem situation. If the proposed changes are accepted by the relevant people as desirable and feasible, the SSM cycle will be completed after performing these changes.

2.3.2   Soft systems methodology: Exercise

Exercise 2.1   Consider the following situation: A professor feels that recently his students are not paying close attention to his lectures. He decides to identify the problems using the SSM. Note the six elements that a root definition should include in reference to CATWOE:

- *Customers*: the beneficiaries or victims of the transformation process

- *Actors*: those who would undertake the transformation process

- *Transformation*: the conversion of input to output

- *Worldview*: the worldview that makes this transformation meaningful

- *Owners*: those who could stop the transformation

- *Environmental constraints*: the elements outside the system that are considered as given

2.4 Oriental Systems Methodologies

In East Asia, systems scientists and practitioners began theorizing their systems methodologies in the 1990s; for a survey and analysis of Oriental systems methodologies, see Zhu (1998). As latecomers, they had the opportunity to use their Western counterparts as a point of reference and departure: what concerns to share, what insights to incorporate, what alienating features to avoid, etc. Such sharing, incorporating, and avoiding is by no means conscious, consistent, or correct; learning is always intuitive. In this section, we have selected the Japanese *Shinayakana*[7] *Systems Approach* and the Chinese *Wuli-Shili-Renli systems approach* as examples that are accessible to Western readers.

2.4.1 Shinayakana Systems Approach

Whatever the reasons, since the last decade of the twentieth century the concepts of intuition and of group collaboration have resulted in novel approaches to knowledge creation, most of which are directly or indirectly related to Japanese origins. Historically, the first of such approaches is the *Shinayakana Systems Approach* by Nakamori and Sawaragi (1990, 1992) in the field of decision and systems science. Being systemic and influenced by the soft and critical systems tradition, it did not specify a process-like, algorithmic recipe for knowledge and technology creation, only a set of principles for systemic problem solving.

 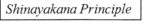

Shinayakana Principle

Using intuition;

Keeping an open mind;

Trying diverse approaches and perspectives;

Being adaptive and ready to learn from mistakes; and

Being elastic like a willow but sharp as a sword – in short, shinayakana

---

[7]Yoshikazu Sawaragi began to use the term *shinayakana approach* before publication. *Shinayakana* is a Japanese word that might correspond to the English word *supple*, but actually it is a word that synthesizes hard and soft. It might be strange for Western people to hear that something is both robust and flexible at the same time.

Here, let us revise hard and soft methodologies.

- Hard systems methodology includes analytical methods of mathematical computerized models of systems problems, including simulation, optimization, multiple criteria analysis, scenario analysis, uncertainty analysis, decision support, etc.

- Soft systems methodology relies on the correct observation that most difficult problems, particularly involving those on the human, soft side, do not have precise mathematical models. If we had such a model, the problem would already be partly solved.

- Since languages are useful codes, but imperfectly describe our deeper knowledge, mathematical language also cannot describe the world perfectly. So the critique of soft systems methodology should not only be applied to hard systems methodology, it should also be applied to itself.

The *Shinayakana Systems Approach* tried to resolve the controversy between hard and soft systems methodologies, using the Eastern philosophy of yin and yang as shown in Figure 2.10.

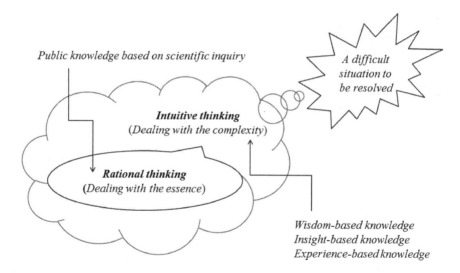

Figure 2.10: Concept of the Shinayakana Systems Approach.

2.4.2 WSR Systems Approach

This section introduces the Chinese Wuli-Shili-Renli (WSR) systems approach (Gu and Zhu, 2000), which is the most accessible to Western people. WSR begins with Confucian folk metaphysics (Nisbett 2003) that conceives human life as a dynamic web of relationships.

- *Wuli* refers to principles or mechanisms related to things, conditions, rules; for instance, natural resources, physical environment, climate, population, transportation and communication means, the funds available, etc.

- *Shili* relates to principles or mechanisms hidden in the pattern of interaction between humans and the world; here we should explore and understand how the world is modeled and managed.

- *Renli* corresponds to principles or mechanisms regarding ethical and social values, interests, etc.; here we should use implicit interest, intentions, or motives of the relevant people in systems design and operation.

The concept of Wuli-Shili-Renli is explained in Figure 2.11. It treats hard approaches in Wuli, both approaches in Shili, and soft approaches in Renli. In this way it uses both approaches complementarily.

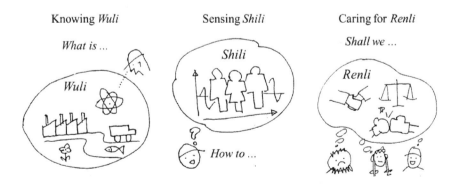

Figure 2.11: The concept of Wuli-Shili-Renli (Gu and Zhu, 2000).

Process   The WSR systems approach is summarized below.[8] See also Figure 2.12.

1. To understand the requirements:

   - Accept, embody, enhance, correct, deform the requests.
   - Tools: brainstorming, soft systems methodology, critical systems thinking

2. To investigate the conditions:

   - Investigate potential resources or funds related to the confirmed request.
   - Tools: questionnaire, the Delphi method, affinity diagram

3. To clarify the purpose:

   - Convert clients' requests into the purpose of the project.
   - Tools: brainstorming, soft systems methodology, decision trees, structural modeling techniques

4. To create a model:

   - Select appropriate models, methods, procedures, and coordination under the agreed objective and conditions identified.
   - Tools: interactive planning, scenario analysis

5. To adjust the relationship:

   - Adjust the subjective relationships between relevant people.
   - Through discussion, new interests or problems emerge.

6. To implement the proposal:

   - Implement the recommended proposal.

7. To evaluate the results:

   - This stage is not the final stage of the project; actually, the project enters a new cycle of improvement.

---

[8]Some systems techniques in the list will be explained in the appendix.

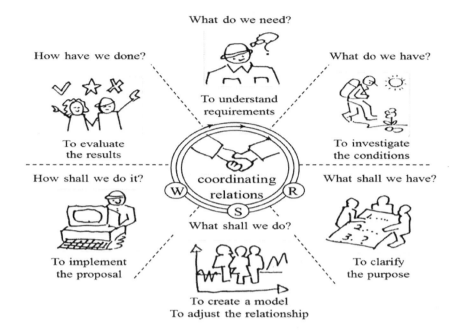

Figure 2.12: The WSR approach (Gu and Zhu, 2000).

Exercise 2.2   Consider an application of the WSR systems methodology, taking an example of a regional revitalization project. The problem is given below:

- *Region*: a small village in a mountainous region

- *Industry*: agriculture and forestry

- *Population*: significant reduction in youth

- *Problems*: increase in fallow fields and uncontrolled forests

- *Possibility*: resort development, hot springs, mountain sports

What will you do in each stage of the WSR approach?[9]

---

[9]For a different version of the WSR process, see Gu (2012).

2.5 Hints for Questions and Exercises

Hint for Question 2.1 Introduce a relationship matrix $R = (r_{ij})$ such that

$$r_{ij} = \begin{cases} 1, & \text{if } j \text{ is used to calculate } i; \\ 0, & \text{if } j \text{ is not used to calculate } i. \end{cases} \tag{2.1}$$

Here, both i and j correspond to Y, C, I, G. Then from the model:

$$\begin{aligned}
Y_t &= C_t + I_t + G_t, \\
C_t &= \alpha Y_{t-1}, \\
I_t &= \beta (C_t - C_{t-1}) + I_{t-1}, \\
G_t &= \gamma Y_{t-1},
\end{aligned}$$

we have

$$R = \begin{pmatrix} 0 & 1 & 1 & 1 \\ 1 & 0 & 0 & 0 \\ 0 & 1 & 1 & 0 \\ 1 & 0 & 0 & 0 \end{pmatrix}. \tag{2.2}$$

Thus, the complexity of a prediction one year in the future is 7. The complexity of a prediction two years ahead is 13 because

$$\begin{pmatrix} 0 & 1 & 1 & 1 \\ 1 & 0 & 0 & 0 \\ 0 & 1 & 1 & 0 \\ 1 & 0 & 0 & 0 \end{pmatrix} \begin{pmatrix} 0 & 1 & 1 & 1 \\ 1 & 0 & 0 & 0 \\ 0 & 1 & 1 & 0 \\ 1 & 0 & 0 & 0 \end{pmatrix} = \begin{pmatrix} 2 & 1 & 1 & 0 \\ 0 & 1 & 1 & 1 \\ 1 & 1 & 1 & 0 \\ 0 & 1 & 1 & 1 \end{pmatrix}. \tag{2.3}$$

In the same manner, when predicting variables five years later and ten years later, the complexities are 75 and 1431, respectively.

Hint for Question 2.2 A *system* as a scientific term, by definition, must meet several requirements for the hierarchical structure, the emergent properties, the functions of communication and control, and the viable potential against a changeable environment. An education system must have a hierarchical structure and emergent properties. It also must have appropriate functions of communication and control. It must have ideas and methods for survival to respond to changes in the environment.

Hint for Question 2.3

1. A system to us is a set of components connected such that properties emerging from it cannot be found in its components. If the whole can be called a system, it has emergent properties that appear as a result of the interaction between its parts.

2. To keep the system viable against a changeable environment, the system must have a higher complexity than the environment. Here, the complexity of a system implies various means, including knowledge and wisdom, which the system has for survival.

Hint for Question 2.4 Complexity includes the complexity of the object itself and the complexity of the human beings involved. In the former case, the number of base elements, the number of relationships between elements, and the number of elements of state all have strong influences on complexity. In the latter case, the interest, abilities, or thinking processes, which are human characteristics, have strong influences on complexity.

1. For the problems that are structured, it is possible to explicitly state the target system. Examples of structured systems are mechanical systems, biological systems, and economic systems. Human elements are not included in these systems. Therefore, the complexity in this case is the complexity of the object itself.

2. For the problems that are not structured, to state the target system explicitly, the system must be simplified. Some examples are management systems, political systems, and social systems. These systems include humans as their elements. Here, system complexity is dramatically increased because of the two types of complexity interacting with each other.

Hint for Question 2.5

1. Consider a situation where you try to grab something. You move your fingers toward the object. You observe the distance between your fingers and the object. You fix the direction of your fingers based on that information. Such operations are the brain's feedback control.

2. The aims of cybernetics and general systems theory:

   - The purpose of cybernetics is to develop a unified theory of control and communication mechanisms with the basic concept of the principle of feedback.

   - General systems theory stresses the appearance of structural similarities or isomorphism in different disciplines, based on the observation that mathematical models of dynamic systems with feedback are the same for mechanics and electronics, and also for biology and sociology, and thus the theory is independent from an actual discipline. Therefore, its aim is to find the isomorphism in different disciplines and to develop a unified, mathematical, trans-disciplinary systems theory.

3. Checkland (1978) defined hard systems thinking as sharing the assumption that the problem task a group tackles is to select an efficient means of achieving a known and defined end.

4. The essential contribution was adding the dimension of human relations, based on the assumption that diverse aspects of human behavior and human relations cannot be adequately modeled mathematically, thus need an essentially different set of systemic procedures.

Hint for Exercise 2.1   A professor feels that recently his students are not paying close attention to his lectures. He decides to identify the problems.

Stage 1

- "I feel that recently my students do not listen carefully to my lectures."

- "I would like to identify the problems using the SSM."

Stage 2   For the students, the lecture should be a system:

- to get the credits necessary for graduation;

- to learn new things useful for a future career; and

- to gain new knowledge useful for research.

Difficulty: Inseparability of knowledge and experience. The professor must deliver knowledge to students without experience.

Stage 3   Root definition: The lecture is a system in which a professor and students co-create new values and new knowledge. See Figure 2.13.

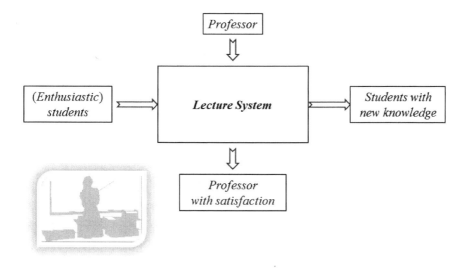

Figure 2.13: An ideal system for students and professors.

Here, the CATWOE can be defined as follows:

- *Customers*: the students

- *Actors*: the professor (plus teaching assistants) and the students

- *Transformation*: The enthusiastic students get new knowledge that is useful for their future.

- *Worldview*: The lecture is a system in which a professor and students co-create new values and new knowledge.

- *Owners*: the professor and the students.

- *Environmental constraints*: There are time constraints. Students must understand new things in a given time.

Stage 4   Conceptual model: An example of the conceptual model to elimi-nate the difficulties for the professor is given in Figure 2.14.

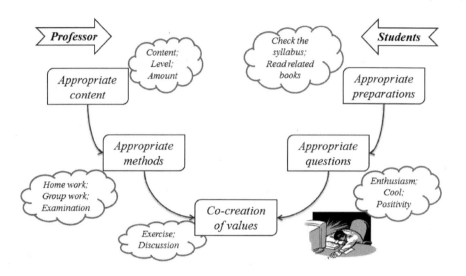

Figure 2.14: A conceptual model to eliminate the difficulties.

Stage 5   All activities in the conceptual model exist and can be judged. Based on the conceptual model, explore solutions for the students who wish to learn new things useful for future careers, or to gain new knowledge useful for research.

Stage 6

1. Inform students of the content of lectures in advance.

2. Ask students to read the related materials in advance.

3. Use user-friendly descriptions to encourage individual thinking.

4. Leave sufficient time for questions, discussion, or exercises.

5. Group work is quite useful to understand theories.

6. Ask students to present their opinions or homework.

Hint for Exercise 2.2 An example of an answer is given below:

	Wuli	*Shili*	*Renli*
Understand requirements	*Desirable economy and society*	*Ideal situations*	*Preferences of residents*
Investigate the conditions	*Possible resources, capitals, funds*	*Possible ways to realize the ideal*	*Cooperation of residents*
Clarify the purpose	*Economic development*	*Discovery of effective methods*	*Settlement of many young people*
Create a model	*Systemically desirable reform*	*Rational ways to achieve the purpose*	*Culturally feasible reform*
Adjust the relationship	*Adjustment between purposes*	*Adjustment between means*	*Adjustment between people*
Implement the proposal	*Revision of goals, if necessary*	*Review of methods, if necessary*	*Sustained cooperation*
Evaluate the results	*Achievement of objectives*	*Effectiveness of the methods*	*Evaluation of cooperation*

Similarity The idea of WSR as a systems methodology is similar to the *i*-System as a knowledge integration model, which will be introduced in the next chapter.

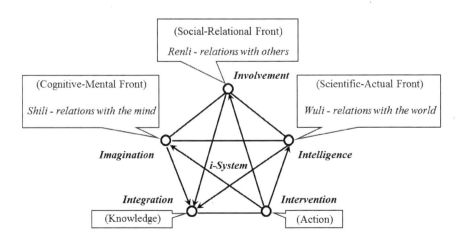

2.6   Appendix on Systems Approaches

Jackson (2003) classifies systems approaches into four types as follows:

Type A: Improving goal seeking and viability

- Hard systems thinking
- System dynamics: the fifth discipline
- Organizational cybernetics
- Complex theory

Type B: Exploring purposes

- Strategic assumption surfacing and testing
- Interactive planning
- Soft systems methodology

Type C: Ensuring fairness

- Critical systems heuristics
- Team syntegrity

Type D: Promoting diversity

- Postmodern systems thinking

The various systems approaches cannot be used all at once, but they can be employed creatively, in an informed and ethical way, to jointly promote the overall improvement of organizational performance. This is the essence of creative holism, and related methodologies are:

- Total systems intervention: the approach to combining different systems approaches

- Critical systems practice: the modern expression of creative holism

This appendix briefly introduces some of these systems approaches, as described in Jackson (2003).

2.6.1 Organizational cybernetics

Cybernetics was originally developed as the science of control and communication in the animal and the machine. It was transferred to the management domain as a hard systems approach initially, where the usefulness of cybernetics was constrained by the machine metaphor (Jackson, 2003).

Organization chart

A person at the top of the organization needs a brain weighing half a ton, since all information must flow up to him and all decisions appear to be his responsibility. People's heads do not get bigger toward the top of an organization.

Beer (1972, 1979, 1985) changed this situation with his *organizational cybernetics*. He redefined cybernetics as the *science of effective organization* and set out to construct a more accurate and useful model called the *viable system model* (VSM), which is a model of the key features that any viable system must exhibit.

Use of the VSM

1. The first step is to agree on the identity of the organization in terms of the purposes it is to pursue. The policy-making function must express and represent these purposes, but obviously should not be the sole repository of identity. Ideally, it should reflect purposes that emerge from and are accepted by the operational elements.

2. The second step involves "unfolding" the complexity of the organization by deciding what operational or business units will enable it to best achieve its purposes. According to the logic of the model, these units will be made as autonomous as possible within the constraints of overall systemic cohesion. The strategy of unfolding complexity, therefore, increases the variety of the organization with respect to its environment.

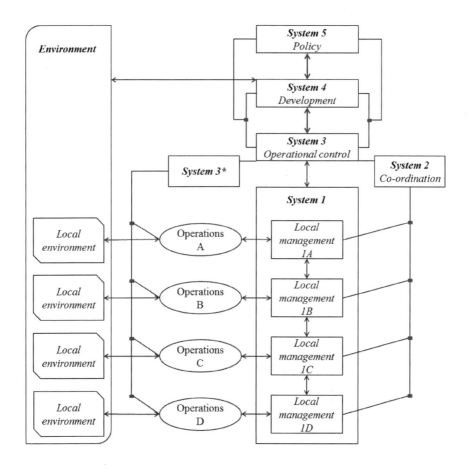

Figure 2.15: The viable system model (VSM).

Viable system model   The VSM[10] is shown in Figure 2.15. It is made up of five elements (System 1 to 5), which we can label implementation, co-ordination, operational control (including service management), development and policy. The functions handled by these five elements must be adequately performed in all systems that wish to remain viable.

---

[10]When the VSM is employed as part of organizational cybernetics, its role is to reveal whether enterprises obey laws or flout them. If used in *design* mode its gaze is focused on plans for new organizations, and it aims to ensure they are constructed according to good cybernetic principles. If used in *diagnostic* mode, it acts as an exemplar of good organization against which the structures and processes of an actually existing system can be checked.

The viable system model consists of five systems, which are explained below (Jackson, 2003):

1. System 1 consists of the various parts of the organization concerned with implementation—with carrying out the tasks directly related to the purposes. In Figure 2.15 the organization has been broken down into four operational elements, labeled A, B, C, and D. Each of these has its own localized management—1A, 1B, 1C, and 1D, and its own relations with the relevant part of the outside world.

2. System 2 consists of the various rules and regulations that ensure the System 1 parts act cohesively and do not get in each other's way. It will also embed any legal requirements that must be obeyed. System 2 is for ensuring harmony between the elements of System 1.

3. System 3* is a servant of System 3, fulfilling an auditing role to ensure that targets specified by System 3 and rules and regulations promulgated by System 2 are being followed. This channel gives System 3 direct access, on a periodic basis, to the state of affairs of the operational elements.

4. The role of System 3 proper is operational control of System 1 and service management of functions such as human resources and finance. It has overall responsibility for the day-to-day running of the enterprise, trying its best to ensure that policy is implemented appropriately.

5. System 4, development, is the place in the organization where internal information received from System 3 is brought together with information about the organization's total environment and presented in a form that facilitates decision making. The primary function of System 4 must be to capture, for the organization, all relevant information about its total environment.

6. System 5, policy, is responsible for the direction of the whole enterprise. It formulates policy on the basis of the information received from System 4 and communicates this downward, to System 3, for implementation by the divisions. An essential task is balancing the often conflicting internal and external demands placed on the organization.

2.6.2 Critical systems heuristics

Ulrich (1983) considers that the systems idea is used only in the context of
instrumental reason to help us to decide how to do things; it refers to a set
of variables to be controlled. His purpose was to develop the systems idea as
part of practical reason, to help us decide what we ought to do. Ulrich calls
his approach *critical systems heuristics*:

- To be *critical* means reflecting on the presuppositions that enter into
  both the search for knowledge and the pursuit of rational action.

- The *systems* idea refers to the totality of elements—ethical, political,
  ideological and metaphysical—on which theoretical or practical judgments depend.

- *Heuristics* refers to the process of continually revealing these presuppositions and keeping them under review.

Ulrich transformed the philosophy and theory he considered attractive into a
methodology applicable to planning and systems design; the result is *critical
systems heuristics*:

1. *Purposeful systems paradigm*: If we wish to understand and improve
   social reality, we must add an additional dimension of *purposefulness*
   and design social systems to become purposeful systems.

2. *Principles for the methodology*: Systems, *moral*, and *guarantor* concepts; quasi-transcendental ideas more suited to social reality and ca-
   pable of acting as critical standards against which the partiality of particular social system designs can be compared.

3. *Boundary judgments*: These reflect the designer's whole system judgments about what is relevant to the design task.

To reveal the boundary judgments involved, boundary questions must be
asked for each of the four groups—client, decision taker, designer and witnesses. Ulrich looks at the nature of the boundary judgments that must inevitably enter into any social systems design. There are 12 boundary judgments in total, which are arranged around a distinction between those involved in any planning decision (client, decision taker, designer) and those
affected but not involved (witnesses). See Table 2.2.

Table 2.2: Ulrich's twelve boundary questions in the *ought* mode.

Involved	Boundary Questions
For client	*Who ought to be the client (beneficiary) of the system S to be designed or improved?*
	What ought to be the purpose of S (i.e., what goal states ought S be able to achieve to serve the client)?
	What ought to be S's measure of success (or improvement)?
For decision taker	*Who ought to be the decision taker (i.e., have the power to change S's measure of improvement)?*
	What components (resources and constraints) of S ought to be controlled by the decision taker?
	What resources and conditions ought to be part of S's environment (i.e., not be controlled by S's decision taker)?
For designer	*Who ought to be involved as designer of S?*
	What kind of expertise ought to flow in to the design of S (i.e., who ought to be considered an expert and what should be his role)?
	Who ought to be the guarantor of S (i.e., where ought the designer seek the guarantee that his design will be implemented and will prove successful, judged by S's measure of success (or improvement)?
For witness	*Who ought to be a witness representing the concerns of the citizens that will or might be affected by the design of S (i.e., who among the affected ought to get involved)?*
	To what degree and in what way ought the affected be given the choice of emancipation from the premises and promises of the involved?
	On what worldview of either the involved or the affected ought S's design be based?

Jackson (2003) summarizes the value of *critical systems heuristics* to managers as follows:

- It offers an inclusive systems approach that emphasizes the benefits of incorporating the values of all stakeholders in planning and decision making.

- It puts the boundary concept at the center of systems thinking and makes it easy to see that drawing the boundary around a problem situation in different ways impacts how it is seen and what is done.

2.6.3   Postmodern systems thinking

Lyotard (1984) recognizes two major manifestations of modernism that can be labeled *systemic modernism* and *critical modernism.*

- *Systemic modernism* is concerned with increasing the performativity of systems, in terms of input–output measures, and with handling environmental uncertainty. It relies on science to discover what is logical and orderly about the world, and on technology to assist with prediction and control. This form of modernism is expressed in classical accounts of hard systems thinking, systems dynamics, organizational cybernetics, and complexity theory.

- *Critical modernism* sees its task as the progressive realization of universal human emancipation. This form of modernism can be recognized in soft systems thinking and in the work of emancipatory systems thinkers.

Postmodernists attack the whole *Enlightenment* rationale, and therefore the pretensions of both systemic and critical modernism. They reject particularly the belief in rationality, truth, and progress. They deny that science can provide access to objective knowledge and so assist with steering organizations and societies in the face of complexity. They deny that language is transparent and can function as a regulative principle of consensus.

- If the postmodernists are right, there are considerable implications for traditional systems thinking. If there is no rationality or optimum solution to problems, then its problem-solving techniques will lack legitimation.

- The fit between postmodernism and the systems thinking we have been studying to date may not look very good. Nevertheless, there are two ways in which systems thinking and postmodernism can collaborate.

  - The first is using various systems methods, models, and techniques, but in the spirit of postmodernism.
  - The second is by postmodernism, providing some new methods and tools that can assist the systems practitioners.

Taket and White (2000) give the name PANDA (Participatory Appraisal of Needs and the Development of Action) to their approach to intervention. PANDA is said to embrace *pragmatic pluralism* and to have grown from postmodern roots. PANDA has four phases and nine tasks to be addressed during the phases, as shown in Table 2.3.

Table 2.3: Four phases and nine tasks in the PANDA.

Phases	Tasks
Deliberation I	*Selecting participants*
	Defining purpose/objectives
	Exploring the situation
Debase	*Identifying options*
	Researching options (which could include consulting on options)
	Comparing options
Decision	*Deciding action*
	Recording decisions
Deliberation II	*Monitoring/Evaluating*

This may look very much like a classical methodology, but Taket and White insist that its application is more an art or craft than a science. In particular, in order to remain true to the spirit of postmodernism, it is essential to recognize and respond to pluralism in each of the four areas:

- in the nature of the client;

- in the use of specific methods;

- in the modes of representation employed; and

- in the facilitation process.

Habermas (1987) recognizes that the postmodernists have something to say, but believes that rather than abandoning the Enlightenment vision, we need to renew and revitalize it. To do this requires more reason—to overcome the difficulties on which postmodernism is focused—rather than less.

2.6.4   Total systems intervention

Total systems intervention (TSI) was heralded as a new approach to planning, designing, problem solving and evaluation based on critical systems thinking (CST). Flood and Jackson (1991) see seven principles as underpinning this meta-methodology:

1. Problem situations are too complicated to understand from one perspective, and the issues they throw up too complex to tackle with quick fixes.

2. Problem situations, and the concerns, issues and problems they embody, should therefore be investigated from a variety of perspectives.

3. Once the major issues and problems have been highlighted, it is necessary to make a suitable choice of systems methodology or methodologies to guide intervention.

4. It is necessary to appreciate the relative strengths and weaknesses of different systems methodologies of the main issues and concerns, to guide the choice of appropriate methodologies.

5. Different perspectives and systems methodologies should be used in a complementary way to highlight and address different aspects of organizations, their issues, and problems.

6. TSI sets out a systemic cycle of inquiry with interaction back and forth between its three phases.

7. Facilitators and participants are engaged at all stages of the TSI process.

The sixth principle refers to the three phases of the TSI meta-methodology, which are labeled *creativity*, *choice*, and *implementation*. This three-phase TSI approach is summarized below:

Creativity

> Task: To highlight significant concerns, issues, and problems
>
> Tools: Creativity-enhancing devices including systems metaphors

Outcome: Dominant and dependent concerns, issues, and problems identified

Choice

Task: To choose an appropriate systems intervention methodology or methodologies

Tools: Methods for revealing the strengths and weaknesses of different systems methodologies (e.g., the SOSM[11])

Outcome: Dominant and dependent methodologies chosen for use

Implementation

Task: To arrive at and implement specific positive change proposals

Tools: Systems methodologies employed according to the logic of TSI

Outcome: Highly relevant and coordinated change that secures significant improvement in the problem situation

Here, some additional notes:

- TSI is a systemic and interactive process. Attention needs to be given during each phase to the likely outcomes of other phases. As the problem situation changes in the eyes of the participants, a new intervention strategy will have to be devised.

- The SOSM is the traditional tool employed by TSI in the choice phase. It unearths the assumptions underlying different systems approaches by asking what each assumes about the system in which it hopes to intervene and about the relationship between the participants associated with the system.

- Combining the information gained about the problem context during the creativity phase and the knowledge provided by the SOSM about the strengths and weaknesses of different systems approaches, it is possible to move toward an informed choice of systems intervention strategy.

---

[11]The system of systems methodologies shown in Figure 2.6.

Adding the dimension of human relations to the complexity dimension in developing systems methodology enables us to facilitate systemic knowledge synthesis.

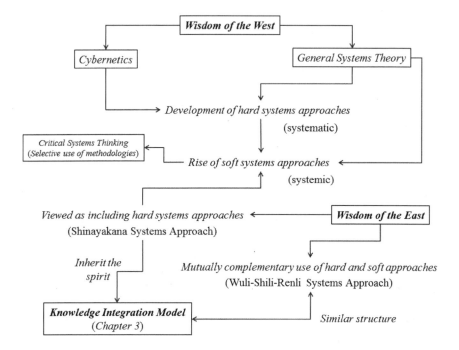

Chapter 3

An Integrated Systems Approach

3.1   Principles of Integration

We have learned about the Oriental systems approaches that are aiming to achieve a synthesis of hard and soft systems approaches while maintaining confrontational complementarity. This chapter, elaborating on these approaches further, introduces the idea of a new systems approach that focuses on intercultural knowledge synthesis. See Figure 3.1.

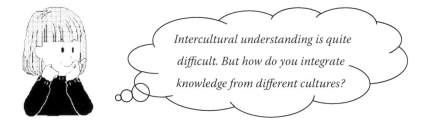

Intercultural understanding is quite difficult. But how do you integrate knowledge from different cultures?

Figure 3.1: How do you achieve intercultural integration?

The challenge of the new systems approach is interdisciplinary and intercultural integration of knowledge, which is a defining feature of new systems science in the knowledge-based society.

Hard and soft systems   How shall we define systems science at the beginning of a new era of knowledge civilization? Before we look at that, let us review the difference between hard and soft systems approaches.

- The mathematical models used in hard systems approaches are intended to be as objective as possible and as exact as needed. Knowledge cannot be absolutely objective and exact; however, hard systems approaches cannot succeed without having objectivity and precision as their goals, while at the same time understanding their limitations.

- The soft systemic approach stresses the intersubjective character of knowledge, the role of disputes, the emancipation of all actors taking part in the intersubjective formation of knowledge, and of a critical approach characterized by three principles: critical awareness, emancipation and human improvement, and pluralism.

We will try to integrate the two approaches, seemingly as different as fire and water.

Perspective of integration   Systems science should be a discipline concerned with methods for the intercultural and interdisciplinary integration of knowledge, including soft intersubjective and hard objective approaches, open, and above all, informed (Wierzbicki, Zhu, and Nakamori, 2006).

- *Intercultural* means an explicit accounting for and analysis of national, regional, even disciplinary cultures, and trying to overcome the incommensurability of cultural perspectives by explicit debate of the different concepts and metaphors used by diverse cultures.

- The *interdisciplinary approach* has been a defining feature of systemic analysis since Comte[1] (1848), but has been gradually lost in the division between soft and hard systems approaches.

- *Open* means pluralist, as stressed by soft systems approaches, not excluding by design any cultural or disciplinary perspectives.

- *Informed* means pluralist as stressed by hard systems approaches, not excluding such perspectives by ignorance or by disciplinary paradigmatic belief.

---

[1]Auguste Comte (1798–1857) is regarded as the originator not only of sociology, but also of modern systemic concepts.

Principles of integration Andrzej P. Wierzbicki declared in Wierzbicki, Zhu, and Nakamori (2006) the following principles of integration with an emphasis on intercultural understanding, which were actually theoretical support for developing the knowledge integration system that will be introduced in Section 3.4. New levels of complexity require new concepts, in a sense transcendental to concepts needed on different levels; transcendental, because they are independent and irreducible to concepts from different levels. This results in:

> The *principle of cultural sovereignty*: We can treat separate levels of systemic complexity as independent cultures, and generalize the principle of cultural anthropology: *no culture shall be judged using concepts from a different culture.*

Apparently, this principle justifies the disciplinary separation of science into diverse fields. Thus, the principle of cultural sovereignty must be accompanied by its dialectic antithesis or yin–yang partner:

> The *principle of informed responsibility*: *No culture is justified in creating a cultural vacuum.* It is the responsibility of each culture to inform other cultures about their own developments and be informed about developments of other cultures.

Again we have a yin–yang relation or a dialectic triad. The above thesis and antithesis must be integrated or synthesized, and we therefore must have a third principle for this purpose:

> The *principle of systemic integration*: Whenever needed, *knowledge from and about diverse cultures and disciplines might be synthesized by systemic methods*, be they soft or hard, without a prior prejudice against any of them, following the principles of open and informed systemic integration.

I understand the principles of integration.

But how can I actually perform them?

3.2   Intercultural Understanding

An even greater challenge, not only for sociology but also for most other fields of science, is not *interdisciplinary* but *intercultural* understanding. But, see Figure 3.2.

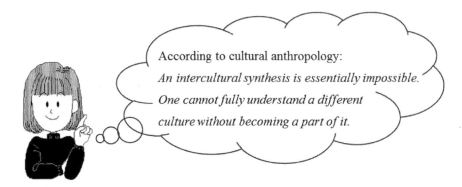

According to cultural anthropology:
An intercultural synthesis is essentially impossible.
One cannot fully understand a different
culture without becoming a part of it.

Figure 3.2: A proposition of cultural anthropology.

3.2.1   Incommensurability

The above proposition was strengthened by the belief of Kuhn (2000) in the incommensurability of different scientific paradigms and thus, by implication, of different cultures:

- Since different paradigms or cultures use different basic concepts, a correct translation of the languages used by these paradigms or cultures is impossible.

On the other hand, we live in a time of *globalization*, which is a necessary condition and at the same time a consequence of the development of global knowledge civilization.

- Either we continue to believe that *intercultural synthesis* is essentially impossible, which will end in one culture dominating others and in a global cultural uniformity, or we revise this belief and find methods of intercultural synthesis and understanding while preserving cultural diversity.

3.2.2 Occidental and Oriental cultures

While we cannot externalize all basic concepts responsible for incommensurability, let us enumerate at least some concepts essential for the mutual understanding of Western and Eastern, Occidental and Oriental cultures. See Figure 3.3.

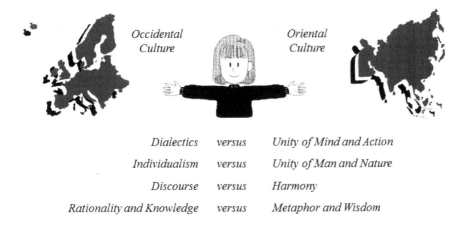

Occidental Culture		*Oriental Culture*
Dialectics	*versus*	*Unity of Mind and Action*
Individualism	*versus*	*Unity of Man and Nature*
Discourse	*versus*	*Harmony*
Rationality and Knowledge	*versus*	*Metaphor and Wisdom*

Figure 3.3: Occidental versus Oriental cultures.

Dialectics vs. Unity of Mind and Action

- The Occidental *dialectic* stresses the value of structured deliberation. Find arguments for a cause in thesis, devise arguments against this in antithesis, and deliberate on synthesis.

- The Oriental concept of *unity of mind and action* is based upon the realization of the essential value of the internalized experience of an expert, and the best practical results are obtained by expert action, not by deliberation.

- In a sense, dialectics is present also in the Oriental principle of the unity of opposed concepts, *yin and yang*, but Oriental thought stresses the essential unity in opposition, while Occidental thought stresses the dialectic process of achieving the unity.

Individualism vs. Unity of Man and Nature

- Occidental *individualism* has a religious foundation in Judaism and Christianity, based on man seeking domination over nature, but also based upon deep philosophic reflection on the moral determination of actions of a human individual, using the Kantian critique of practical reason.

- The Oriental principle of *unity of man and nature*, related to the principle of the unity of opposed concepts, also has a very long tradition in religious beliefs, such as Buddhism.

- Both perspectives have their advantages and disadvantages: the danger of polluting the natural environment results from extreme *individualism*, while the danger of excusing inaction and stasis stems from extreme interpretations of the unity of man and nature, as in some aspects of Taoist doctrine.

Discourse vs. Harmony

- Occidental societies, by stressing social and legal confrontations, adhere more to the principle of *discourse* or *debate* that attaches bonuses to winning rational arguments.

- The Oriental principle of *harmony* stems in some sense from the principle of the unity of man and nature: nature is single, thus unity of man and nature implies the harmonious actions of many individuals.

  – But it is also ingrained in the Oriental character by Confucian doctrine that an individual has many obligations, to his/her family, to his/her teachers, to the group to which he/she belongs; thus he/she should seek consensus and harmony with others.

  – This principle has many advantageous consequences: it makes Oriental societies more stable and more able to achieve consensus than Occidental societies. However, it also has disadvantages: a major one is that consensus is often achieved only tacitly, intuitively, without a sufficiently deep rational examination of possible options.

- These principles are not directly opposed. In fact, Occidental societies might learn methods of achieving consensus from Oriental ones. On the other hand, Oriental societies might improve their debating skills by learning from Occidental ones.

    - Oriental, in particular Japanese society, has often been accused of *collectivism*, which is a misunderstanding; competitive behavior is quite natural to Japanese. *Harmony* means much more than collectivism, though it implies some aspects of collectivist behavior. Therefore, we discuss harmony in opposition to Occidental discourse rather than individualism.

Rationality and Knowledge vs. Metaphor and Wisdom

- Occidental principles of *rationality and knowledge* are also related to those of discourse and debate.

    - The use of these two concepts, *rationality and knowledge*, has greatly supported Occidental collective actions in developing industrial civilization. For example, they were the essence of Occidental knowledge that was brought to Japan after the Meiji Restoration.[2]

- Oriental societies attributed historically less importance to these principles, and more to *metaphor and wisdom*.

    - *Metaphor* is the intuitive and emotional synthesis of a specific piece of knowledge; *wisdom* is a post-rational attribute of an expert or sage, a deep but again intuitive synthesis of general knowledge. The doctrines of Tao and Buddhist Zen philosophy use metaphors widely and stress that true *wisdom* is achieved by forgetting the rational prejudices of an expert.

---

[2]The Meiji Restoration was a chain of events that restored imperial rule to Japan in 1868, which led to enormous changes in Japan's political and social structure. The leaders of the Meiji Restoration acted in the name of restoring imperial rule in order to strengthen Japan against the threat represented by the colonial powers of the day. The word *Meiji* means enlightened rule and the goal was to combine Western advances with the traditional Eastern values.

3.2.3   Discussion of intercultural understanding

Question 3.1

1. Discuss globalization and cultural diversity.

2. Consider intercultural understanding and synthesis.

Hint

1. *Viewpoints for discussion*:

   - Some people believe that the problem of multiple cultures will solve itself in a kind of cultural evolution: globalization forces will simply impose the dominant culture on other nationalities or cultures.

   - There is, however, a big danger related to such a course: global cultural uniformity without informed understanding of our diversified cultural legacy will increase the threat of global conflicts.

   - Therefore, preserving cultural diversity is very important, especially in a time of globalization. This follows from Ashby's *Law of Requisite Variety* (Ashby, 1958), but today is of particular significance.

2. *Viewpoints for discussion*:

   - We can expect serious conflicts in the knowledge civilization era. Each national or regional culture, preserved but well informed and understood in the context of global civilization, increases our chances of solving unpredictable problems related to these inevitable conflicts.

   - Thus, intercultural synthesis should not aim to promote cultural uniformity, but to increase understanding about diverse cultures. With this aim, we must analyze the limits of incommensurability.

   - We have many examples of the actual incommensurability of diverse natural languages, but people speaking those languages still manage to communicate and to develop global knowledge, science, technology, and a global heritage of humanity.

3.3 An Informed Systems Approach

The new goal of systems science is to integrate soft and hard systems approaches, rather than setting them against each other, and to provide for a better understanding between West and East, instead of assuming that they will never meet. The future diversity of human civilization and its robustness in the face of unexpected challenges depends on such understanding; for that purpose, we stress the importance of intercultural integration of knowledge (Wierzbicki, Zhu, and Nakamori, 2006). See Figure 3.4.

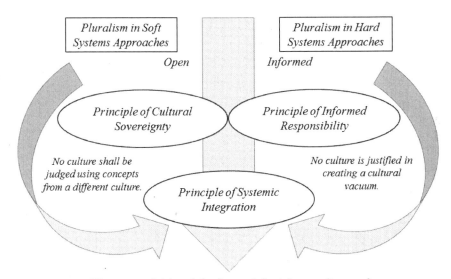

Whenever needed, knowledge from and about diverse cultures and disciplines can be synthesized by systemic integration.

Interdisciplinary, Intercultural Knowledge Integration

Figure 3.4: An Informed Systems Approach.

This integration must also be interdisciplinary; it must include soft intersubjective and hard objective approaches, and it must be open and informed. Informed means not neglecting inconvenient information, and implies trying to understand the viewpoint of a different discipline without paradigmatic prejudice; it means knowing, like a true master, all pertinent approaches and methods, and using them selectively, elastically, and incisively.

Knowledge integration   A new interpretation of systems science includes a new understanding of necessary and unnecessary reduction, of synergy and the value of complementarity relations, of the emergence of new systemic properties on further levels of complexity, and of pluralism and systemic multiplicity.

- It does not assume the abolishment of differences between soft and hard systems approaches.

- It welcomes productive competition between these sub-disciplines.

- But it requires mutual respect and informed attitudes, instead of misinformed criticism.

- Its main aim is integration, in particular, *knowledge integration*. See Figure 3.5.

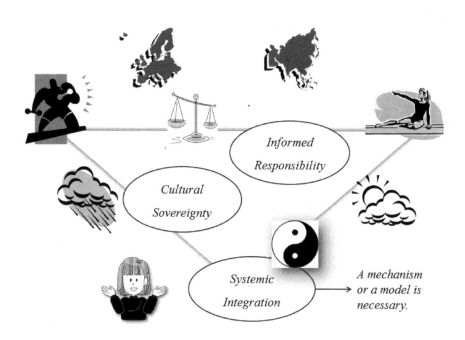

Figure 3.5: The main aim is to integrate knowledge from various sources.

3.4 A Knowledge Integration System

A knowledge integration system called the *i*-System, proposed in Nakamori (2000, 2003), is a procedural (but virtually systemic) approach to knowledge creation. In fact, it is a knowledge integration model representative of the *Informed Systems Approach*. See Figure 3.6. The five ontological elements or subsystems of the *i*-System are:

- *Intervention* (the will to solve problems)

- *Intelligence* (existing scientific knowledge)

- *Involvement* (social motivation)

- *Imagination* (other aspects of creativity)

- *Integration* (systemic knowledge)

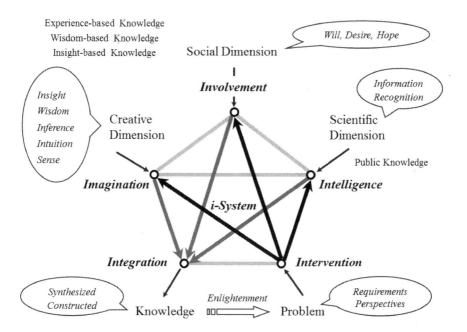

Figure 3.6: A knowledge integration system (the *i*-System).

3.4.1   The *i*-System

The five ontological elements were originally interpreted as nodes, as illustrated in Figure 3.6.

- Since the *i*-System is intended as a synthesis of systemic approaches, *Integration* is, in a sense, its final dimension.

- In Figure 3.6 all arrows converge on *Integration*, one of the nodes; links without arrows denote the possibility of impact in both directions.

- The beginning node is *Intervention*, where problems or issues perceived by individuals or the group encourage further inquiry to start the entire creative process.

- The *Intelligence* node corresponds to various types of knowledge.

- The *Involvement* node represents social aspects.

- The creative aspects are represented mostly in the *Imagination* node.

Further interpretations

- The *i*-System defines three important dimensions and requires various actors to work well in each dimension in collecting and organizing distributed, tacit knowledge. These are *Intelligence* (scientific dimension), *Involvement* (social dimension), and *Imagination* (creative dimension).

- The *i*-System requires two more dimensions, *Intervention* and *Integration*, which possibly correspond to Midgley's *systemic intervention*.[3]

- The *i*-System aims at integrating a *systematic approach* and *systemic (holistic) thinking*.[4] The former is mainly used in the dimensions of *Intelligence*, *Involvement*, and *Imagination*, and the latter is required in the dimensions of *Intervention* and *Integration*.

---

[3]Systemic intervention in Midgley (2000, 2004) is purposeful action by an agent to create change in relation to reflection upon boundaries.

[4]Leading systems thinkers today often emphasize *holistic thinking* (Jackson, 2003; Mulej, 2007), or *meta-synthesis* (Gu and Tang, 2005).

Connections with the Informed Systems Approach   The *i*-System is regarded as a concrete methodology of an *Informed Systems Approach*, and characterized as follows:

- Hard pluralism in *Intelligence* (according to the principle of informed responsibility)

  - Tries to examine a variety of methods and tools without prior prejudice against them, based on a belief in a certain disciplinary paradigm.

  - Tries to treat not only rational knowledge, but also intuitive knowledge, and further, emotional knowledge.

- Soft pluralism in *Involvement* (according to the principle of cultural sovereignty)

  - Tries to take into account any cultural or disciplinary standpoints.

  - Tries to have a sense of perspective for the benefits to individual, group, and society.

- Holistic thinking in *Imagination* (use of experience- and insight-based knowledge, wisdom)

  - Tries to imagine and simulate the future based on a variety of information.

  - Tries to stimulate idea generation using brainstorming or information techniques.

However, the *i*-System does not clearly offer an answer to the question "How can we perform systemic knowledge integration?" Instead, the *i*-System emphasizes:

- repetition of *Intervention* and *Integration* for systemic intervention, and

- involvement of *knowledge coordinators* as elements of the system to achieve the principle of systemic integration.

3.4.2   Discussion of characteristics of the *i*-System

Question 3.2

1. Discuss the features of the *i*-System, a knowledge integration system. The features include:

   - Fusion of the purposefulness and purposiveness paradigms.
   - Interaction of explicit knowledge and tacit knowledge.
   - Involvement of knowledge coordinators.

2. Consider the necessary agencies of knowledge coordinators.

Hint   *Fusion of the paradigms*: With the *i*-System we always start by searching for and defining the problem following the purposefulness (generating a purpose) paradigm. Since the *i*-System is a spiral-type knowledge integration model, in the second cycle we use the *i*-System to find solutions following the purposiveness (achieving the purpose) paradigm. However, it is almost always the case that just when we have found an approximate solution, we face new problems. By introducing the viewpoint of knowledge construction, the *i*-System enables us to use the purposefulness and purposiveness paradigms simultaneously. See Figure 3.7.

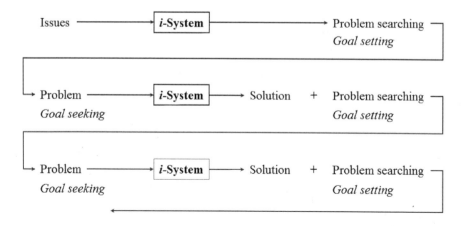

Figure 3.7: Fusion of the purposefulness and purposiveness paradigms.

1. Problem solving begins with a desire to collect knowledge. Let us denote this knowledge as A.

2. The first use of the i-System is to investigate what kinds of knowledge are actually available. Actors usually obtain integrated knowledge, which might consist of explicit intellectual assets, implicit or tacit knowledge among people, and knowledge that actors already have. We can denote this integrated knowledge as B.

3. Almost always, there is a difference between A and B. There is probably some knowledge that actors cannot obtain despite their best efforts. The difference between A and B is a function of the amount of knowledge one can obtain in the time available.

4. Actors have to create new knowledge, C, to fill in the gap between A and B. The i-System cannot know C in advance. The creation of C is the work of actors. If the creation of C is difficult, actors have to restart their search for it with the i-System.

Hint *Interaction of explicit and tacit knowledge*: This book accepts the idea of Nonaka and Takeuchi (1995) that new knowledge can be obtained by the interaction between explicit knowledge and tacit knowledge. The use of the i-System means that we have to inevitably integrate objective knowledge, such as scientific theories, available technologies, and socioeconomic trends, as well as subjective knowledge, such as experience, technical skills, hidden assumptions, and paradigms, for a certain purpose. See Figure 3.8.

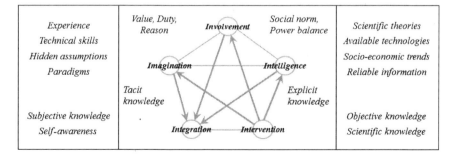

Figure 3.8: Interaction of explicit and tacit knowledge.

Hint   *Knowledge coordinators*: The system requires people who will accomplish the synthesis of knowledge. Such persons need to have the abilities of knowledge workers and of innovators in wide-ranging areas. However, they cannot achieve satisfactory results unless they possess the ability to coordinate the opinions of people and diverse knowledge. We should establish an education system to train people who will promote knowledge synthesis in a systemic manner.

- The necessary abilities in the three dimensions:

    - *Intelligence*: Ability to understand information properly
    - *Involvement*: Ability to communicate with others properly
    - *Imagination*: Ability to imagine the future properly

- In addition, the knowledge coordinator must also have these abilities:

    - *Intervention*: Ability to define a problem properly
    - *Integration*: Ability to synthesize knowledge properly

3.4.3   Types of integration

Integration in the original *i*-System is a node intended to represent the final stage, the systemic synthesis of the creative process. Thus, in this stage we should use all systemic knowledge; application of systemic concepts to newly created knowledge is certainly the only explicit, rational knowledge tool that can be used to achieve integration. Thus, any teaching of creative abilities must include a strong component of systems science.

- The simplest level is *specialized integration*, in which the task consists of integrating several elements of knowledge in some specialized field.

- The task becomes more complex if its character is *interdisciplinary*, as in the case of the analysis of environmental policy models.

- However, the contemporary trends of globalization result in new, even more complex challenges related to *intercultural integration*, as in the case of integration of diverse theories of knowledge and technology creation.

Specialized integration   The example shown in Figure 3.9 relates to the specialized knowledge integration process in a project for the development of traditional translucent porcelain in Japan.

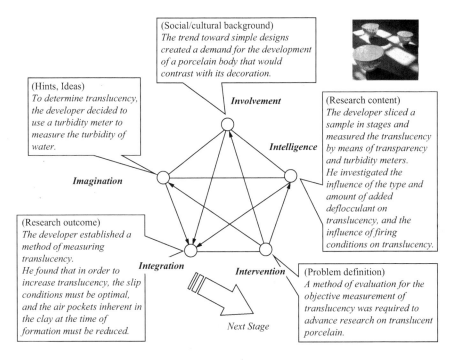

Figure 3.9: Research and development of translucent porcelain.

We collected the following knowledge, created in recent technology development projects in that industry (Yamashita and Nakamori, 2007):

Intervention: motivation (objective, mission), external pressure (needs, requests), problem definition (subject, direction)

Intelligence: research content (methods, tools), research facilities, existing research (scientific knowledge, literature)

Involvement: budget, industrial conditions, collaboration (organizations, enterprises, individuals), social/cultural backgrounds (socioeconomic conditions, trends)

Imagination: hints and ideas (enlightenment), hidden stories (problems, failures), attitudes (an ideal, interests, enthusiasm, changes of heart)

Integration: research results (content, results, discoveries), understanding (success or failure, new problems), evaluation (self-judgment, external evaluation), commodification (production, sale, patent), new research plans (for the next stage of the project)

Interdisciplinary integration   With the goal of developing a fresh-food management system, we tried to integrate both a hard systematic approach and a soft systemic approach (Ryoke et al., 2007). See Figure 3.10.

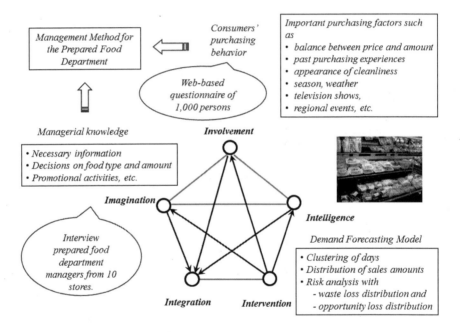

Figure 3.10: A fresh-food management system.

Intervention: To deal with a demand prediction problem, a new approach was necessary to integrate the system engineering approach and the knowledge management approach. We decided to collect knowledge about facts, social relations, and personal recognition in the dimensions of *Intelligence*, *Involvement*, and *Imagination*.

Intelligence: We constructed a demand prediction model based on past sales data, using data mining techniques, clustering analysis, etc., and developed a system for risk management to cope with waste loss as well as opportunity loss.

Involvement: We collected and analyzed consumers' opinions, using a web-based questionnaire, to investigate important factors that affect purchasing decisions, which cannot be found from the point-of-sale data or from the opinions of managers.

Imagination: We collected managerial knowledge from some competent managers, which included factors regarding decisions on types and amounts of goods, and decisions regarding advertising promotions, taking into account special circumstances.

Integration: We developed a management system consisting of a prediction subsystem, a risk analysis subsystem, and a managerial subsystem. However, the final decision should be made by a competent manager, who has experience and intuition, based on the output from the prediction and risk analysis system.

Intercultural integration   When establishing a plan for regional revitalization, taking into account environment, healthcare, biomass, etc., we have to consult authorities such as scientists, politicians, and economists, as well as civil servants, and most importantly, local residents. This requires interdisciplinary and intercultural integration of knowledge. See Figure 3.11. Here, we return to the problem of promoting a biomass town plan discussed in Section 1.3.

Intervention: Let us consider the promotion of a biomass town plan. Suppose that the problem is to find reasons for slow progress in establishing a biomass town.

Intelligence: Investigate the existing scientific knowledge and the obstacles for promoting the biomass town plan, by interviewing researchers and administrative officials.

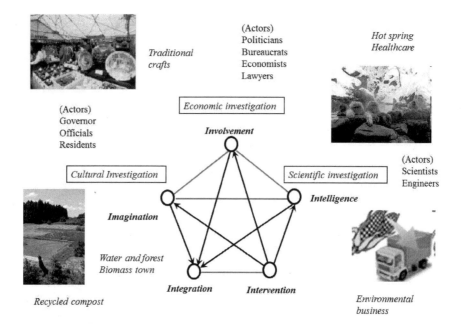

Figure 3.11: Planning of regional revitalization.

Involvement: Investigate the views of people who are already cooperating with this plan, asking them about the results so far, and their evaluation of the results.

Imagination: Investigate citizen awareness of environmental problems, asking them what they know about the biomass town plan, how they learned about it, whether they are willing to cooperate with the plan, and why or why not.

Integration: Assemble knowledge from the above three subsystems and summarize the problems in promotion of the biomass town plan.

Applications of specialized integration, interdisciplinary integration, and intercultural integration will be given in Section 3.5, Section 7.2, and Section 8.5, respectively.

3.5 Application to Technology Archiving

This section presents an application of the *i*-System to technology archiving for a traditional craft industry in Japan. The results of this study indicated that the *i*-System is useful for passing down a rich variety of knowledge on technical development to future generations, and would thereby support future research and development.

3.5.1 Knowledge collection

Besides a contribution to the regional economy, traditional craft industries have a high cultural value, as they involve techniques that are unique to individual regions. While the slump in the traditional craft industry continues, there are signs of a trend toward preserving works with highly artistic value in the form of digital data. But up until now, there has not been much research on the preservation of technical innovations in the traditional craft industry. Also, there is no database system for distributing knowledge to the technology developers who support the traditional craft industry behind the scenes.

Traditional crafts The traditional craft industry treated here is Ishikawa Prefecture's Kutani-ware (ceramics) industry. Ishikawa Prefecture has the Kutani-ware Research Center, a public organization that specializes in giving technical support to the Kutani-ware industry. This center keeps precise reports on technical development, which helps us to collect objective knowledge. In addition, we interviewed technical developers to collect subjective or even implicit knowledge, following the idea of the *i*-System, which limited our survey to techniques developed in the last 30 to 40 years.

Three dimensions

- *Intelligence*: Technical innovations in the Kutani-ware industry were investigated. Business reports published by the Kutani-ware Research Center were used as the main source of information.

- *Involvement*: To obtain knowledge about individual technical innovations, previous research and case examples were investigated through the use of existing literature and interviews with technical developers. The main focus in this investigation was the relationship with social and cultural background.

- *Imagination*: People who had been involved in technical development for many years were interviewed in detail. They talked about what they had focused on while engaged in technical development, and shared their ideas and thoughts. This enabled us to collect the implicit knowledge required to carry out technical innovations.

Interview survey

- The subjects of the survey were eight people involved with technical development in the Kutani-ware industry, who belonged to public organizations in Ishikawa Prefecture.

- The interview started with basic questions pertaining to their personal history (work experience, job description, specialization), the motivation for starting their research, information on production areas, and the typical flow of the research.

- The interview continued with questions focused on the essence of their research (methods for solving problems, the situation of collaborators and cooperating organizations, the influence of research results on production areas, and thoughts about the research and changes in attitudes at various stages).

- Questions tailored to each research project were added to these basic questions.

Systematization   The interview data was classified and systematized according to the *i*-System. As for the specific method of systematization, a timeline for the technology development process was set as follows:

1. Research motivation and background

2. Research and development

3. Practical applications

4. Commercialization

5. Improvement

The collected data for each stage was classified into five dimensions: *Intervention*, *Imagination*, *Involvement*, *Intelligence*, and *Integration*. To facilitate classification and organization, we used expressions that were as specific as possible, as shown in Figure 3.12.

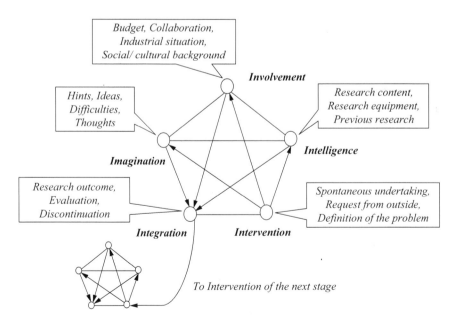

Figure 3.12: Specific expressions used for interview.

3.5.2   Knowledge archiving

Collecting and systematizing knowledge pertaining to the Kutani-ware industry based on the *i*-System enabled us to determine things that were previously known only by people involved with technical innovation. Such information includes cultural and economic conditions, research processes, cooperation with companies in the production area and public organizations, flashes of inspiration that led to solutions to problems, and changes in thoughts and feelings about the research being carried out. Combining such knowledge with existing information, which we obtained from research reports, makes it possible to gain a deeper understanding of past technical innovations, and will undoubtedly contribute to future technical development. Thus, systematizing information and knowledge according to the *i*-System strongly supports research and development activities.

A brief summary of knowledge collected based on the *i*-System is presented in Figure 3.13 and the subsequent figures.

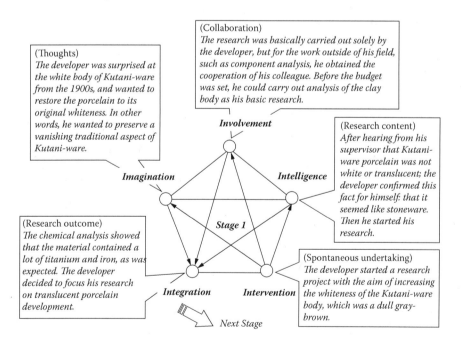

Figure 3.13: Motivation for development of translucent porcelain.

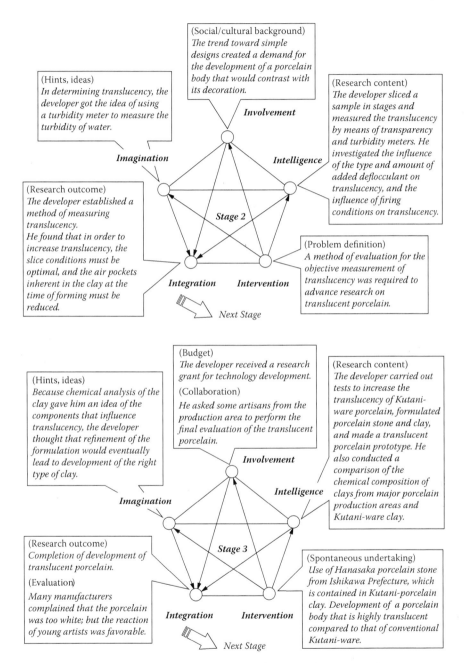

Figure 3.14: Research and development (two stages).

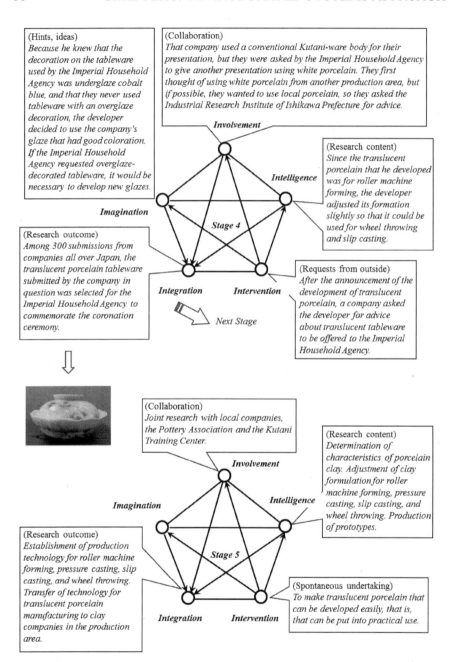

(Hints, ideas)
Because he knew that the decoration on the tableware used by the Imperial Household Agency was underglaze cobalt blue, and that they never used tableware with an overglaze decoration, the developer decided to use the company's glaze that had good coloration. If the Imperial Household Agency requested overglaze-decorated tableware, it would be necessary to develop new glazes.

(Collaboration)
That company used a conventional Kutani-ware body for their presentation, but they were asked by the Imperial Household Agency to give another presentation using white porcelain. They first thought of using white porcelain from another production area, but if possible, they wanted to use local porcelain, so they asked the Industrial Research Institute of Ishikawa Prefecture for advice.

(Research content)
Since the translucent porcelain that he developed was for roller machine forming, the developer adjusted its formation slightly so that it could be used for wheel throwing and slip casting.

(Research outcome)
Among 300 submissions from companies all over Japan, the translucent porcelain tableware submitted by the company in question was selected for the Imperial Household Agency to commemorate the coronation ceremony.

(Requests from outside)
After the announcement of the development of translucent porcelain, a company asked the developer for advice about translucent tableware to be offered to the Imperial Household Agency.

Involvement, Intelligence, Imagination, Integration, Intervention, Stage 4, Next Stage

(Collaboration)
Joint research with local companies, the Pottery Association and the Kutani Training Center.

(Research content)
Determination of characteristics of porcelain clay. Adjustment of clay formulation for roller machine forming, pressure casting, slip casting, and wheel throwing. Production of prototypes.

(Research outcome)
Establishment of production technology for roller machine forming, pressure casting, slip casting, and wheel throwing. Transfer of technology for translucent porcelain manufacturing to clay companies in the production area.

(Spontaneous undertaking)
To make translucent porcelain that can be developed easily, that is, that can be put into practical use.

Involvement, Intelligence, Imagination, Integration, Intervention, Stage 5

Figure 3.15: Practical applications and commercialization.

3.6 Application to Technology Roadmapping

The *i*-System can be used for constructing roadmaps[5] in the field of technology management.

Intervention An *intervention* can be understood as a motivational dimension, a drive, or determination, or even dedication to solving a problem. Starting a roadmapping process can be thus thought of as an intervention for issues motivating strategic plans.

1. First, initiators of the roadmapping process should have a deep understanding of what the motivation is for making the particular roadmap.

2. Second, they should understand what roadmaps and roadmapping are, the advantages of roadmapping, and how to do roadmapping.

3. Third, initiators or coordinators must also consider who should participate in the roadmapping team and encourage them to join, customize a roadmapping process and schedule, and let all participants know the purpose and schedule and their roles in the roadmapping.

Intelligence *Intelligence* has two aspects: *rational or explicit* and *intuitive or tacit*. It is a duty of the coordinator and of all participants in the roadmapping process to search for relevant explicit information. In this task, the following tools could be helpful:

- Scientific databases: Access to disciplinary or general scientific databases such as Scopus, ScienceDirect, etc., can be very helpful for researchers to understand what has been done, what is being done, and what needs to be done.

- Text mining tools: The amount of scientific literature increases very quickly; thus, help in finding relevant explicit information is necessary.

---

[5]Robert Galvin, the former CEO of Motorola, defined technology roadmapping as a vision for future research and action, or the making of a plan enhanced by a vision.

- Workshops: Hold workshops in which many experts are involved. Here, some specific groupware, such as Pathmaker, could be applied to structure and manage discussions among experts.

In fact, the third tool already involves some elements of intuitive or tacit knowledge from experts. But an important aspect of obtaining good intelligence is individual reflection on and interpretation of the explicit information previously obtained.

Involvement   This is a social dimension, related to two aspects: societal motivation and consensus building in the group of participants. Roadmapping in a group is a consensus-building process. This process might include many researchers, experts, and other stakeholders. This dimension includes the following important aspects:

- Participation of administrative authorities: If administrative authorities are involved in the coordination of the roadmapping process, then the process proceeds more smoothly.

- Customized solutions: Preparing a template for the roadmapping process also helps it to proceed smoothly. There are many existing solutions that could serve as templates, such as the T-plan (Phaal et al., 2001), disruptive technology roadmaps (Kostoff et al., 2004), and interactive planning solutions for personal research roadmaps (Ma et al., 2005).

- Internet-based groupware: The use of Internet-based groupware can contribute to *Involvement*.

Imagination   *Imagination* is needed during the entire roadmapping process; it should help to create vision. Participants are encouraged to imagine a purposeful future and the means to get there.

- Graphical presentation tools: Graphical presentation tools can help people to express and refine their thoughts.

- Simulations: Simulations can enhance and stimulate imagination, especially concerning complex dynamic processes.

- Critical debate: This is probably the most fundamental way of encouraging imaginative thinking.

- Brainstorming: Brainstorming is, in a sense, a counterpart of critical debate; it encourages people to generate and express diverse, perhaps even brilliant ideas, and is directly related to imaginative thinking.

- Idealized design: Idealized design is a unique and essential feature of the *Interactive Planning* approach (Ackoff, 1974, 1978, 1981), which is regarded as a basic method for solving creative problems.

Integration *Integration* must be applied several times during the process of roadmapping, at the very least, when making a draft, a revised version, and the final version of a roadmap. *Integration* includes all knowledge of the other four dimensions, and is therefore interdisciplinary and systemic. Diverse rational systemic approaches may be helpful:

- Analytical hierarchy process (Saaty, 1980, 1990) (See Section 3.7.1)

- Meta-synthesis approach[6] (Gu and Tang, 2005)

However, in order to be creative and visionary, *Integration* cannot rely only on rational, explicit knowledge, but must rely on preverbal, intuitive, and emotional knowledge. Therefore, software with a heuristic interface and graphical representation tools are essential for help in this dimension.

- For example, there might be numerous nodes and links in a roadmap that would be difficult to master by an unaided human brain.

- A properly chosen graphical representation of the roadmap might thus be essential.

- In order to choose such a representation, a heuristic interface can be applied to infer the preferred features of graphical roadmaps.

---

[6]The meta-synthesis approach was proposed by a Chinese systems scientist, Xuesen Qian, around the early 1990s. It tackled open, complex, and giant system problems, which could not be effectively solved by traditional reductionism methods. The method emphasizes the synthesis of collected information and knowledge of various experts, and combining quantitative methods with qualitative knowledge.

3.7   Appendix on Systems Techniques

Among many systems techniques we here introduce the *Analytical Hierarchy Process* and *Interpretive Structural Modeling.*

3.7.1   The Analytical Hierarchy Process

The *Analytical Hierarchy Process* (AHP) (Saaty, 1980) is one of the decision-making methods that enables a person to select the most advantageous alternative for his/her objective. As shown in Figure 3.16, we arrange the goal, evaluation criteria, and alternatives hierarchically.

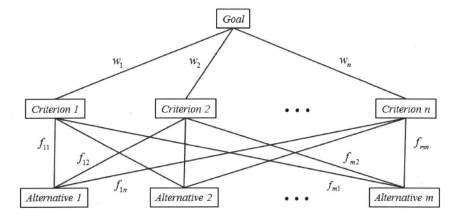

Figure 3.16: Hierarchical diagram for evaluation of alternatives.

Let $f_{k1}, f_{k2}, \cdots, f_{kn}$ be the scores of criteria for the alternative k, and let $w_1, w_2, \cdots, w_n$ be the importance of criteria as seen from the goal. The overall rating value of the alternative k is calculated by a linear combination:

$$T_k = w_1 f_{k1} + w_2 f_{k2} + \cdots + w_n f_{kn}, \quad k = 1, 2, \cdots, m.$$

In Saaty's method, there is a unique idea to determine the importance of criteria $w_1, w_2, \cdots, w_n$.

Pairwise comparison   Define an $n \times n$ matrix $A = (a_{ij})$ by the following pairwise comparison:

How important is the criterion i compared with the criterion j?

Here, the elements of A are determined by

$$a_{ij} = \begin{cases} 1, & \text{Equivalent;} \\ 3, & \text{Somewhat important;} \\ 5, & \text{Important;} \\ 7, & \text{Considerably important;} \\ 9, & \text{Extremely important;} \end{cases}$$

$$a_{ii} = 1, \quad a_{ji} = \frac{1}{a_{ij}}, \quad i,j = 1,2,\cdots,n.$$

Estimation of importance Since a_{ij} is regarded as an estimate of w_i/w_j, in the ideal case, the pairwise comparison matrix becomes

$$A = \begin{pmatrix} w_1/w_1 & w_1/w_2 & \cdots & w_1/w_n \\ w_2/w_1 & w_2/w_2 & \cdots & w_2/w_n \\ \vdots & \vdots & \ddots & \vdots \\ w_n/w_1 & w_n/w_2 & \cdots & w_n/w_n \end{pmatrix}.$$

If we denote the important vector by

$$\boldsymbol{w} = (w_1, w_2, \cdots, w_n)^t \quad (t \text{ represents the transpose}),$$

we have

$$A\boldsymbol{w} = n\boldsymbol{w}.$$

That is, n is an eigenvalue of A, and $\boldsymbol{w}$ is the eigenvector corresponding to n. Note that n is the maximum eigenvalue of A. In general, it is difficult to obtain such an ideal matrix. Therefore, letting the maximum eigenvalue and the corresponding eigenvector be

$$\lambda_{\max}, \quad \boldsymbol{v} = (v_1, v_2, \cdots, v_n)^t,$$

we estimate the importance of the item i by

$$w_i = \frac{v_i}{\displaystyle\sum_{j=1}^{n} v_j}, \quad i = 1, 2, \cdots, n.$$

Consistency   As an indicator to examine the consistency of the pairwise comparison, the consistency index

$$CI = \frac{\lambda_{\max} - n}{n - 1}$$

is proposed. It is said that we have to redo the pairwise comparison if

$$CI > 0.15.$$

General notes

- Select criteria that are independent.

- Ascertain the consistency of the pairwise comparison.

- Do not give positive meanings to the differences between the overall rating values.

3.7.2   The Interpretive Structural Modeling

Many structural modeling techniques have been proposed; see for instance Lendaris (1980). Here, *Interpretive Structural Modeling* (Warfield 1974, 1976) is outlined. This modeling technique consists of the following steps:

1. Extract the elements of the system.

2. Introduce a perspective (a binary relation) for structuring the object as a system.

3. Perform the pairwise comparison of elements based on the introduced perspective.

4. Investigate the conflicts and indirect relationships in a pairwise comparison.

5. Explore the hierarchical structure of elements to extract the system structure.

Binary relation   Denote a set S consisting of m elements by

$$S = \{s_1, s_2, \cdots, s_m\}.$$

The set of all ordered pairs (s_i, s_j) is called the Cartesian product of S:

$$S \times S = \{(s_i, s_j) \mid s_i, s_j \in S\}.$$

A binary relation R on S is a subset of $S \times S$:

$$(s_i, s_j) \in R \iff s_i \text{ is in the relationship of } R \text{ to } s_j.$$

Adjacency matrix   The binary relation R is represented by the adjacency matrix $A = (a_{ij})$, which is an $m \times m$ matrix whose row and column names are $s_1, s_2, \cdots, s_m$, defined by

$$a_{ij} = \begin{cases} 1, & (s_i, s_j) \in R; \\ \\ 0, & (s_i, s_j) \notin R. \end{cases}$$

Digraph   The binary relation R is also represented by the digraph

$$G = (S, R)$$

in which the elements of S are defined as vertices of the graph, and an arrow is drawn from s_i to s_j if $(s_i, s_j) \in R$. An example is given in Figure 3.17.

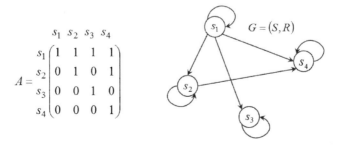

$$R = \{(s_1, s_1), (s_1, s_2), (s_1, s_3), (s_1, s_4), (s_2, s_2), (s_2, s_4), (s_3, s_3), (s_4, s_4)\}$$

$$A = \begin{array}{c} \\ s_1 \\ s_2 \\ s_3 \\ s_4 \end{array} \begin{array}{cccc} s_1 & s_2 & s_3 & s_4 \\ \begin{pmatrix} 1 & 1 & 1 & 1 \\ 0 & 1 & 0 & 1 \\ 0 & 0 & 1 & 0 \\ 0 & 0 & 0 & 1 \end{pmatrix} \end{array}$$

Figure 3.17: An example of the binary relation.

Properties of relations The following are properties of binary relations:

- Reflective law $\Leftrightarrow (x, x) \in R$

- Symmetric law $\Leftrightarrow (x, y) \in R \rightarrow (y, x) \in R$

- Asymmetric law $\Leftrightarrow (x, y) \in R, (y, x) \in R \rightarrow x = y$

- Transitive law $\Leftrightarrow (x, y) \in R, (y, z) \in R \rightarrow (x, z) \in R$

For the system structure, the following two relations are important:

- A binary relation that satisfies the reflective, symmetric, and transitive laws is called an *equivalence relation*.

- A binary relation that satisfies the reflective, asymmetric, and transitive laws is called a *partial order*.

Reachability matrix When the relation R does not satisfy the transitive relation, *interpretive structural modeling* seeks the smallest transitive relation including R. The corresponding matrix is called the *reachability matrix*. Under the Boolean operation

$$0 + 0 = 0, \quad 0 + 1 = 1, \quad 1 + 0 = 1, \quad 1 + 1 = 1,$$

a reachability matrix $M = (m_{ij})$ satisfies

$$M^2 = M, \quad M + I = M,$$

where I is the $m \times m$ identity matrix. The reachability matrix is obtained by

$$M = I + A + A^2 + \cdots + A^k = (I + A)^k, \quad {}^{\exists}k < m.$$

An example is given below:

$$A = \begin{pmatrix} 0 & 1 & 0 & 0 \\ 0 & 0 & 1 & 0 \\ 0 & 0 & 0 & 1 \\ 0 & 0 & 0 & 0 \end{pmatrix} \Rightarrow M = (I + A)^3 = \begin{pmatrix} 1 & 1 & 1 & 1 \\ 0 & 1 & 1 & 1 \\ 0 & 0 & 1 & 1 \\ 0 & 0 & 0 & 1 \end{pmatrix}.$$

Structural modeling Starting with a reachability matrix, we extract the structure using the following steps:

1. *Division into parts*: Divide elements into independent parts, and change the reachability matrix to a block diagonal matrix.

2. *Division into levels*: Divide elements in each part into hierarchical levels, and change each block of the block diagonal matrix to be a lower triangle matrix.

3. *Division in levels*: Divide elements in each level into groups that include interrelated elements, or irrelevant elements.

4. *Extraction of a skeleton*: Draw a digraph with the minimum number of arrows: a structural model.

Division into parts Let the reachability set of s_i be

$$R(s_i) = \{s_j \in S \mid m_{ij} = 1\},$$

and let the antecedent set of s_i be

$$A(s_i) = \{s_j \in S \mid m_{ji} = 1\}.$$

The set of elements consisting of the roots of the graph is given by

$$T = \{s_i \in S \mid R(s_i) \cap A(s_i) = A(s_i)\}.$$

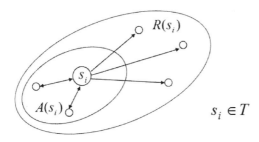

Then, we can find the parts of the graph by the following rules:

- $u, v \in T$, $R(u) \cap R(v) \neq \phi \rightarrow u, v$ belong to the same part.

- $u, v \in T$, $R(u) \cap R(v) = \phi \rightarrow u, v$ belong to different parts.

Division into levels   Assume that a part P is divided into levels:

$$L_1 \text{ (the top level)}, L_2, \cdots$$

The elements of each level are found by the following algorithm:

Step 1: $L_0 = \phi$, $l = 1$.

Step 2: $L_l = \{s_i \in Q_{l-1} \mid R_{l-1}(s_i) \cap A_{l-1}(s_i) = R_{l-1}(s_i)\}$, where

$$Q_{l-1} = P - L_0 - L_1 - \cdots - L_{l-1},$$

$$R_{l-1}(s_i) = \{s_j \in Q_{l-1} \mid m_{ij} = 1\},$$

$$A_{l-1}(s_i) = \{s_j \in Q_{l-1} \mid m_{ji} = 1\}.$$

Step 3: If $Q_l = \phi$, stop; else let $l = l + 1$, and go to Step 2.

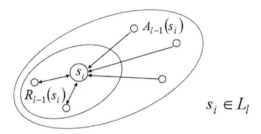

$$s_i \in L_l$$

Division within levels   Let us assume the level L is divided into groups:

$$U_1, U_2, \cdots,$$

- $u, v \in U_i \Leftrightarrow (u, v) \in R, (v, u) \in R$

- $u \in U_i, v \in U_j \ (i \neq j) \Leftrightarrow (u, v) \notin R, (v, u) \notin R$

Choose one element as a representative of the group, and consider the group as one element. The matrix obtained by this operation is called a *reduction matrix*.

Extraction of a skeleton   Find a skeleton matrix A' from the reduction matrix M'. Find the skeleton matrix with the minimum number of 1's among the matrices, the reachability matrix of which corresponds to the reduction matrix; the corresponding graph is the structural model.

Exercise 3.1 Consider the set of seven elements

$$S = \{s_1, s_2, \cdots, s_7\}$$

and a reachability matrix

$$M = \begin{pmatrix} 1 & 0 & 0 & 0 & 0 & 0 & 0 \\ 1 & 1 & 0 & 0 & 0 & 0 & 0 \\ 0 & 0 & 1 & 1 & 1 & 1 & 0 \\ 0 & 0 & 0 & 1 & 1 & 1 & 0 \\ 0 & 0 & 0 & 0 & 1 & 0 & 0 \\ 0 & 0 & 0 & 1 & 1 & 1 & 0 \\ 1 & 1 & 0 & 0 & 0 & 0 & 1 \end{pmatrix}.$$

1. Find the parts of the graph.

2. Find the levels of each part.

3. Find the groups in each level of each part.

4. Extract the skeleton and draw the structural model.

Hint

Skeleton matrix

$$A' = \begin{matrix} & \begin{matrix} s_5 & s_4 & s_3 & s_1 & s_2 & s_7 \end{matrix} \\ \begin{matrix} s_5 \\ s_4 \\ s_3 \\ s_1 \\ s_2 \\ s_7 \end{matrix} & \begin{pmatrix} 0 & 0 & 0 & 0 & 0 & 0 \\ 1 & 0 & 0 & 0 & 0 & 0 \\ 0 & 1 & 0 & 0 & 0 & 0 \\ 0 & 0 & 0 & 0 & 0 & 0 \\ 0 & 0 & 0 & 1 & 0 & 0 \\ 0 & 0 & 0 & 0 & 1 & 0 \end{pmatrix} \end{matrix}$$

$$\{s_4, s_6\} \rightarrow s_4$$

Part 1   Part 2

The perspective of knowledge justification in specialized integration includes novelty, usefulness, and logicality, in the respective dimensions, as with the evaluation criteria of a scientific paper.

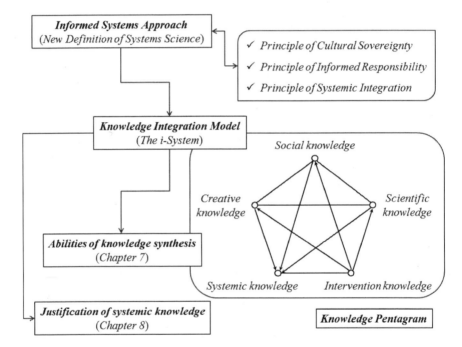

Chapter 4

Mathematical Information Aggregation

4.1 Complexity in Decision Making

Life is a continuous decision-making process, although we know it is full of contradictions. Multicriteria decision analysis is a discipline that has been devoted to solving complex decision-making problems. However, it is still difficult to reconcile the value of various attributes. See Figure 4.1.

I want a ceramic cup that is

✓ *Cute and smooth; and quite modern;*

✓ *Largish; but not too expensive;*

✓ *The most suitable gift for my grandmother.*

Figure 4.1: How could you help her?

Systemic knowledge synthesis If you have been involved in the sale of ceramics for a long time, say more than 20 years, you will immediately recommend several ceramic cups that fit her desired attributes.

> *This is clearly an example of systemic knowledge synthesis.*

Figure 4.2 shows an approach to systemic knowledge synthesis. In an actual situation, the shop owner, synthesizing knowledge from the scientific dimension, the social dimension, and the creative dimension, would recommend some cups to this customer. But in this chapter we will study how to create analytical knowledge mainly in the scientific dimension.

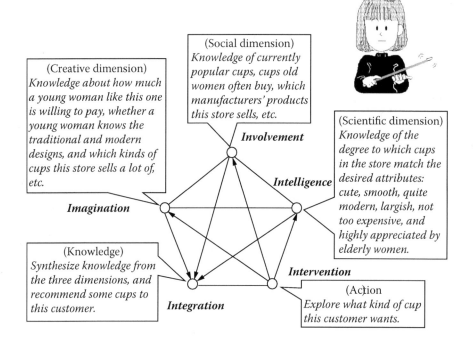

Figure 4.2: An approach to systemic knowledge synthesis.

Multiple criteria Recall the desired attributes of the customer in Figure 4.1, which are classified into three types of information. See Figure 4.3.

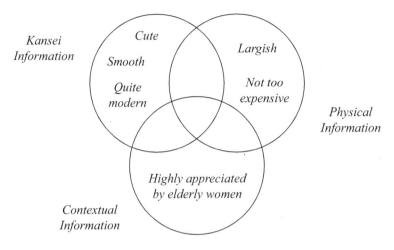

Figure 4.3: Three types of information.

Unified framework Although these are different types of information, and have been dealt with mainly in Kansei (Affective) Engineering,[1] Decision Science, and Knowledge Engineering, respectively, this chapter deals with them in an unified framework. See Figure 4.4:

	-3 -2 -1 0 +1 +2 +3	
Cute	□ ■ □ □ □ □ □	*Refined*
Smooth	□ □ □ □ □ ■ □	*Rough*
Traditional	□ □ □ □ □ □ ■	*Modern*
Large	□ □ ■ □ □ □ □	*Small*
Cheap	□ □ □ □ ■ □ □	*Expensive*
For seniors	■ □ □ □ □ □ □	*For youth*

Figure 4.4: A unified framework to deal with desired attributes.

---

[1] *Kansei* is a Japanese word, meaning the recognition of individual subjective impressions from a certain artifact, environment or situation using all senses of sight, hearing, feeling, smell, and taste. *Kansei Engineering* is a discipline that deals with consumers' subjective impressions and images of a product (artifact and service) and how they are incorporated into design elements. See Nagamachi (2011), Nagamachi and Lokman (2011).

Intelligence and Integration   This chapter is intended to create complete knowledge of *Intelligence* and partial knowledge of *Integration* as in Figure 4.2. This will be done by:

- Creating knowledge of the degree of alignment between cups in the store and respective desired attributes such as cute, smooth, largish, etc. (Intelligence);

- Creating knowledge of the overall ranking of cups in the store by aggregating respective degrees of alignment (Integration).

Knowledge synthesis at *Integration* is actually done by taking into account knowledge from *Involvement* and *Imagination*. Here, let us consider partial tasks at *Intervention*, *Intelligence*, and *Integration* as shown in Figure 4.5.

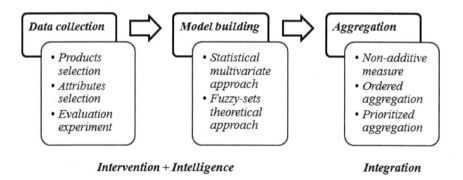

Figure 4.5: The task flow for solving the problem.

The rest of this chapter consists of the following:

- Data collection and screening

    - Evaluation experiment and selection of attributes and data

- Calculation of the degree of fit between requests and products

    - Statistical and fuzzy-set theoretical data processing

- Methods of information aggregation

    - Three types of aggregation operators

4.2 Data Collection and Screening

To build a model of the alignment between desired attributes and products, we need to collect a good data set. See Figure 4.6.

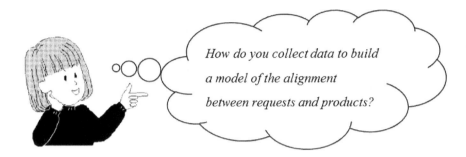

How do you collect data to build a model of the alignment between requests and products?

Figure 4.6: How do you collect and screen data to build a model?

Procedure for data collection and screening

1. *Preparation*: The first task is to prepare products for evaluation by the bipolar measures used in the *semantic differential method* (Osgood et al., 1957), which is often used in subjective evaluation experiments. Here, the most difficult task is to choose words, mainly adjectives, to be used in the measure.

2. *Experiment*: The evaluation experiment should be designed carefully. Depending on the products to be evaluated, we have to gather appropriate evaluators, and teach them the purpose of the experiment, how to do the scoring, etc.

3. *Screening*: The data screening is sometimes necessary because of errors or biased scoring. Moreover, those who are familiar with the products and those who do not know them would make different scores in some measures. Therefore, we need to select data as well as appropriate bipolar measures (examples will be shown later) before going into the modeling phase.

4.2.1   Evaluation experiment

The evaluation of products is performed by using bipolar measures, a pair of words, for instance

$$\langle\text{cool, warm}\rangle.$$

The product attributes are evaluated on a scale of seven grades:

{very cool, cool, slightly cool, neutral, slightly warm, warm, very warm}.

Using the following numbers, we quantify these seven levels:

$$\{-3, -2, -1, 0, +1, +2, +3\}.$$

See Figure 4.7.

Figure 4.7: Products are evaluated by bipolar measures.

Measures   Figure 4.8 shows an example of a set of words for evaluation, which includes measures for kansei attributes and contextual attributes. In fact, to choose appropriate words is not an easy task.

	-3 -2 -1 0 +1 +2 +3	
Smooth	☐☐☐☐☐☐☐	Rough
Cool	☐☐☐☐☐☐☐	Warm
Busy	☐☐☐☐☐☐☐	Plain
Candid	☐☐☐☐☐☐☐	Shy
Luxurious	☐☐☐☐☐☐☐	Simple
Calm	☐☐☐☐☐☐☐	Excitable
Cute	☐☐☐☐☐☐☐	Refined
Plain	☐☐☐☐☐☐☐	Flashy
Light	☐☐☐☐☐☐☐	Heavy
Lively	☐☐☐☐☐☐☐	Serene
Gentle	☐☐☐☐☐☐☐	Strong
Dynamic	☐☐☐☐☐☐☐	Static
Rural	☐☐☐☐☐☐☐	Urban
Delicate	☐☐☐☐☐☐☐	Robust
Fresh	☐☐☐☐☐☐☐	Old
Sociable	☐☐☐☐☐☐☐	Reserved
Traditional	☐☐☐☐☐☐☐	Modern
Feminine	☐☐☐☐☐☐☐	Masculine
Dignified	☐☐☐☐☐☐☐	Casual
Naive	☐☐☐☐☐☐☐	Intelligent
For seniors	☐☐☐☐☐☐☐	For young people
For females	☐☐☐☐☐☐☐	For males
For Western-style rooms	☐☐☐☐☐☐☐	For Japanese-style rooms
For personal use	☐☐☐☐☐☐☐	For a gift
For visitors' use	☐☐☐☐☐☐☐	For routine use
As a souvenir	☐☐☐☐☐☐☐	As a gift

Figure 4.8: Bipolar measures for kansei evaluation.

4.2.2   Data screening

It is difficult for an ordinary person to judge whether a cup is traditional or modern. In fact, different opinion distributions were obtained from experts and consumers. See Figure 4.9, in which we can see the opinion distributions when evaluating a cup by the measure $\langle traditional,\ modern \rangle$.

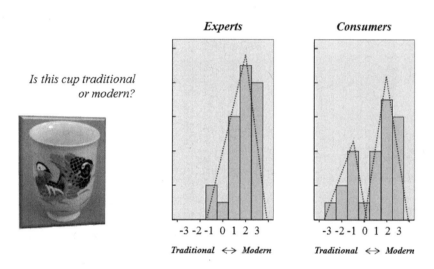

Figure 4.9: Different opinion distributions between experts and consumers.

We see from Figure 4.9 that general consumers do not know precisely the difference between traditional works and modern works. In such a case we have to decide whether we do not use the pair $\langle traditional,\ modern \rangle$ at all or whether we use the data reflecting the experts' opinions only. In addition to this, we must carefully select data:

- Since too many criteria are cumbersome for the user, we must choose a relatively small number of criteria in an actual evaluation system, by using factor analysis or correlation analysis.

- In addition to the classification of experts and consumers, we must deal with data looking at the differences by gender or age. Sometimes, we need to create plural conditional models by splitting the data.

4.2.3 Mathematical data description

Let us introduce the following notations:

- Objects of evaluation: o_m, $m = 1, 2, \cdots, M$

- Measures of evaluation: w_n, $n = 1, 2, \cdots, N$

- Evaluators: e_k, $k = 1, 2, \cdots, K$

Thus, we need to handle *three-mode structure data.*

Data A measure of evaluation w_n is given by a pair of bipolar words:[2]

$$w_n : \langle w_n^-, w_n^+ \rangle, \quad w_n^- : \text{left word}, \quad w_n^+ : \text{right word}.$$

The evaluated value z_{mnk} of the evaluator e_k, regarding the object o_m, from the standpoint of w_n is given by a $(2L + 1)$-level value:

$$z_{mnk} \in \{-L, \cdots, 0, \cdots, L\}, \quad L : \text{a positive integer}.$$

Note that as extreme cases we have

$$z_{mnk} = \begin{cases} -L, & w_n^- \text{ is strongly supported;} \\ L, & w_n^+ \text{ is strongly supported.} \end{cases}$$

This is an extremely vague value, so it is often referred to as *kansei* (affective) data. Generally speaking, in many cases the kansei evaluation data might not be a complete three-mode set. For instance, some evaluators might skip the evaluation of some objects or forget to fill in some measures of evaluation. Therefore, we need to give the data structure a realistic basis.

A pair of bipolar words: < smooth, rough >

A scale of seven grades:

{very smooth, smooth, a little smooth, neutral, a little rough, rough, very rough}

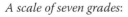

---

[2]A measure of evaluation w_n will be called a word hereafter, unless there is confusion.

Realistic setting Let E be the set of all evaluators, and O be the set of all objects of evaluation. Then, letting E_m be the set of evaluators who evaluate the object o_m, and O_k be the set of objects evaluated by the evaluator e_k with at least one measure of evaluation, we assume

$$E = \bigcup_{m=1}^{M} E_m; \quad E_m \neq \phi, \; {}^{\forall}m;$$

$$O = \bigcup_{k=1}^{K} O_k; \quad O_k \neq \phi, \; {}^{\forall}k.$$

To consider the possibility of missing values, we denote the set of evaluators who evaluate the object o_m with the measure of evaluation w_n by E_{mn}. Then,

$$E_{mn} \subseteq E_m, \; {}^{\forall}n; \quad E_m = \bigcup_{n=1}^{N} E_{mn}.$$

Here, we assume that E_{mn} is not empty for all m and n.

Special cases As special cases, we encounter the following:

Case 1: The case where all objects are evaluated by all evaluators:

$$E_m = E, \; {}^{\forall}m \quad (O_k = O, \; {}^{\forall}k).$$

Case 2: The case where only one object is evaluated by each evaluator:

$$E_m \bigcap E_{m'} = \phi, \text{ if } m \neq m'; \quad E = \sum_{m=1}^{M} E_m.$$

Examples of these cases are given below:

- When we collect a large group of people together and ask them to evaluate some products from several perspectives, we will usually get a *three-mode structure* data set.

- When we conduct a questionnaire survey on the evaluation of residential environments, we usually ask each respondent to evaluate only the environment around him/her.

The concrete example in this chapter corresponds to Case 1. However, we often find data that cannot be used in the analysis due to the errors of respondents. For this reason, we introduce the set E_{mn}.

4.2.4   Data matrices

Here we prepare organized data matrices for the subsequent sections, which are the average data matrix and the frequency matrices.

Average data matrix

$$
Z = \begin{pmatrix}
z_{11} & z_{12} & \cdots & z_{1N} \\
z_{21} & z_{22} & \cdots & z_{2N} \\
\vdots & \vdots & \ddots & \vdots \\
z_{M1} & z_{M2} & \cdots & z_{MN}
\end{pmatrix}. \tag{4.1}
$$

Here, assuming that every E_{mn} is not empty, we put[3]

$$
z_{mn} = \frac{1}{|E_{mn}|} \sum_{k \in E_{mn}} z_{mnk}.
$$

Frequency matrices

$$
Y_n = \begin{pmatrix}
y_{1n(-L)} & y_{1n(-L+1)} & \cdots & y_{1nL} \\
y_{2n(-L)} & y_{2n(-L+1)} & \cdots & y_{2nL} \\
\vdots & \vdots & \ddots & \vdots \\
y_{Mn(-L)} & y_{Mn(-L+1)} & \cdots & y_{MnL}
\end{pmatrix} \tag{4.2}
$$

for $n = 1, 2, \cdots, N$. The elements of Y_n are given by

$$
y_{mnl} = |\{e_k \in E_{mn}; z_{mnk} = l\}|, \; l \in \{-L, \cdots, 0, \cdots, L\}.
$$

That is, y_{mnl} is the number of evaluators who gave the level l to the object o_m from standpoint w_n.

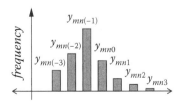

---

[3] $|E_{mn}|$ indicates the number of elements in the set E_{mn}.

4.3   Degree of Fit between Requests and Products

The problem here is how to calculate the degree of fit between requests and products. See Figure 4.10.

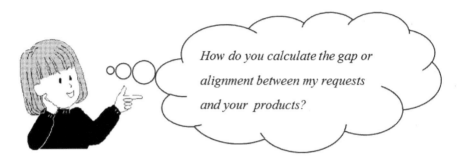

Figure 4.10: The alignment between requests and products.

Three approaches   We consider three approaches to data processing:

1. *Statistical approach*: Among many statistical approaches, factor analysis is mostly used to obtain information about the gaps between objects, between words, and between objects and words. But here, we introduce correspondence analysis, which measures the gaps between objects and words directly.[4]

2. *Probabilistic approach*: To treat the degree of the requirement, such as "a little cute" or "quite traditional" without making any models, we can use the frequencies obtained from the data directly. We can convert the frequencies to probabilities; for instance, the degree of "a little cute" of this cup is given by a certain probability.

3. *Fuzzy-set theoretical approach*: The above approach is acceptable if we have a plenty of data. In the usual examples, we develop models that interpolate data distributions; Gaussian-type probabilistic models are often used. But here, taking into account the possibility of data, we use the triangular fuzzy (possibility) model.

---

[4]A fuzzy version of correspondence analysis is given in Nakamori and Ryoke (2006).

4.3.1 Statistical data processing

Here we shall consider the handling of kansei data using correspondence analysis (Hirschfeld, 1935).

Correlation matrix We first normalize the average data $\{z_{mn}\}$ in (4.1) as follows. If we focus on the words to the right in Figure 4.8, let

$$z'_{mn} = (L+1) + z_{mn} \in [1, 2L+1].$$

On the contrary, if we focus on the words to the left, let

$$z'_{mn} = (L+1) - z_{mn} \in [1, 2L+1].$$

Then, letting

$$p_{mn} = \frac{z'_{mn}}{z_T}, \quad z_T = \sum_{m=1}^{M} \sum_{n=1}^{N} z'_{mn},$$

we define a correlation matrix:

$$P = \begin{pmatrix} p_{11} & p_{12} & \cdots & p_{1N} \\ p_{21} & p_{22} & \cdots & p_{2N} \\ \vdots & \vdots & \ddots & \vdots \\ p_{M1} & p_{M2} & \cdots & p_{MN} \end{pmatrix}.$$

Note that if we let

$$p_{m\bullet} = \sum_{n=1}^{N} p_{mn}, \quad p_{\bullet n} = \sum_{m=1}^{M} p_{mn},$$

the following holds:

$$\sum_{m=1}^{M} p_{m\bullet} = \sum_{n=1}^{N} p_{\bullet n} = 1.$$

Note 4.1 As a special case of correspondence analysis, there is a method that uses the 0-1 matrix by defining the elements of P, for instance,

$$p_{mn} = \begin{cases} 1, & \text{if } z'_{mn} \geq L+1; \\ 0, & \text{if } z'_{mn} < L+1. \end{cases}$$

Objective of correspondence analysis In correspondence analysis, assuming that

- a quantity x_m is associated with the object o_m, and

- a quantity y_n is associated with the word w_n,

we will find the following vectors, which maximize the correlation coefficient ρ_{xy} to be defined below:

$$
\boldsymbol{x} = \begin{pmatrix} x_1 \\ x_2 \\ \vdots \\ x_M \end{pmatrix}, \quad
\boldsymbol{y} = \begin{pmatrix} y_1 \\ y_2 \\ \vdots \\ y_N \end{pmatrix}.
$$

Correlation coefficient The variance and covariance of $\boldsymbol{x}$ and $\boldsymbol{y}$ are given by

$$
\sigma_x^2 = \sum_{m=1}^{M} p_{m\bullet} x_m^2 - \left(\sum_{m=1}^{M} p_{m\bullet} x_m \right)^2,
$$

$$
\sigma_y^2 = \sum_{n=1}^{N} p_{\bullet n} y_n^2 - \left(\sum_{n=1}^{N} p_{\bullet n} y_n \right)^2,
$$

$$
\sigma_{xy} = \sum_{m=1}^{M} \sum_{n=1}^{N} p_{mn} x_m y_n - \left(\sum_{m=1}^{M} p_{m\bullet} x_m \right) \left(\sum_{n=1}^{N} p_{\bullet n} y_n \right).
$$

The correlation coefficient of $\boldsymbol{x}$ and $\boldsymbol{y}$, to be maximized, is given by

$$
\rho_{xy} = \frac{\sigma_{xy}}{\sigma_x \sigma_y}.
$$

The purpose of correspondence analysis is to give similar weightings to objects and words that have similar reactions.

Note 4.2 In the usual scenarios, the mean and variance are calculated with equal weightings $1/M$, but here they are calculated by using different weights:

$$
\bar{x} = \frac{1}{M} \sum_{m=1}^{M} x_m \;\rightarrow\; \sum_{m=1}^{M} p_{m\bullet} x_m,
$$

$$\sigma_x^2 = \frac{1}{M} \sum_{m=1}^{M} (x_m - \bar{x})^2 \rightarrow \sum_{m=1}^{M} p_{m\bullet} (x_m - \bar{x})^2.$$

Eigenvalue problem   The solution for the maximization problem satisfies the following equations:

$$\frac{\partial \rho_{xy}}{\partial x} = 0, \quad \frac{\partial \rho_{xy}}{\partial y} = 0.$$

From this we have

$$P\tilde{y} - \rho_{xy} P_x \tilde{x} = 0, \tag{4.3}$$

$$P^t \tilde{x} - \rho_{xy} P_y \tilde{y} = 0. \tag{4.4}$$

Here, P_x and P_y are the following diagonal matrices:

$$P_x = \begin{pmatrix} p_{1\bullet} & 0 & \cdots & 0 \\ 0 & p_{2\bullet} & \cdots & 0 \\ \vdots & \vdots & \ddots & \vdots \\ 0 & 0 & \cdots & p_{M\bullet} \end{pmatrix},$$

$$P_y = \begin{pmatrix} p_{\bullet 1} & 0 & \cdots & 0 \\ 0 & p_{\bullet 2} & \cdots & 0 \\ \vdots & \vdots & \ddots & \vdots \\ 0 & 0 & \cdots & p_{\bullet N} \end{pmatrix}.$$

And, $\tilde{x}$ and $\tilde{y}$ are the column vectors consisting of elements that are standardized as follows:

$$\tilde{x}_m = \frac{1}{\sigma_x} \left(x_m - \sum_{m=1}^{M} p_{m\bullet} x_m \right),$$

$$\tilde{y}_n = \frac{1}{\sigma_y} \left(y_n - \sum_{n=1}^{N} p_{\bullet n} y_n \right).$$

The following eigenvalue problem is derived from (4.3) and (4.4):

$$Ae = \lambda e. \tag{4.5}$$

Here, we set

$$A = P_x^{-\frac{1}{2}} P P_y^{-1} P^t P_x^{-\frac{1}{2}} \quad \lambda = \rho_{xy}^2, \quad e = P_x^{\frac{1}{2}} \tilde{x}.$$

By solving this eigenvalue problem, we have

$$\tilde{x} = P_x^{-\frac{1}{2}} e, \quad \tilde{y} = \lambda^{-\frac{1}{2}} P_y^{-1} P^t \tilde{x}.$$

Note that the maximum eigenvalue ($\lambda = 1$) is a meaningless solution.

Note 4.3   The following eigenvalue problem is derived if we first erase $\tilde{x}$:

$$\left(P_y^{-\frac{1}{2}} P^t P_x^{-1} P P_y^{-\frac{1}{2}} - \rho_{xy}^2 I \right) P_y^{\frac{1}{2}} \tilde{y} = 0.$$

Graphical display   We obtain eigenvalues $\lambda_1, \lambda_2, \cdots, \lambda_M$ such that

$$1 = \lambda_1 \geq \lambda_2 \geq \cdots \geq \lambda_M \geq 0.$$

Let $\tilde{x}_i$ and $\tilde{y}_i$, which are derived from the eigenvectors corresponding to λ_2 and λ_3, be

$$\tilde{x}_i = \begin{pmatrix} \tilde{x}_{i1} \\ \tilde{x}_{i2} \\ \vdots \\ \tilde{x}_{iM} \end{pmatrix}, \quad \tilde{y}_i = \begin{pmatrix} \tilde{y}_{i1} \\ \tilde{y}_{i2} \\ \vdots \\ \tilde{y}_{iN} \end{pmatrix}, \quad i = 2, 3.$$

From this we plot

$$(\tilde{x}_{2m}, \tilde{x}_{3m}), \quad m = 1, 2, \cdots, M,$$

$$(\tilde{y}_{2n}, \tilde{y}_{3n}), \quad n = 1, 2, \cdots, N,$$

on a plane such as in Figure 4.11, and try to understand the relationships between objects and words. Here, it is desirable to give appropriate meanings to the axes.

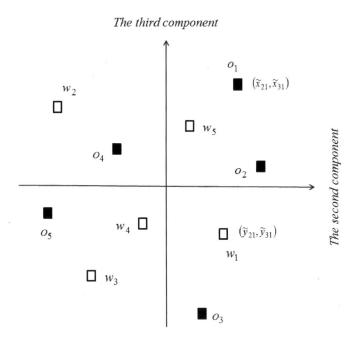

Figure 4.11: An example of the correspondence analysis.

Alignment between objects and words   The gap d_{mn} between the object o_m and the word w_n can be calculated by

$$d_{mn}^2 = (\tilde{x}_{2m} - \tilde{y}_{2n})^2 + (\tilde{x}_{3m} - \tilde{y}_{3n})^2.$$

Then we can define the alignment s_{mn} between the object o_m and the word w_n by

$$s_{mn} = \exp\{-d_{mn}\}. \tag{4.6}$$

In later sections, we will use this value as a measure of the degree of fit between the object o_m and the word w_n when a customer wants a product that is represented by a word w_n.

Note 4.4   Since we have come to this point using the words to the right w_n^+, let us assume a degree of alignment or fit between the object o_m and the word w_n^- as $1 - s_{mn}$.

Alignment matrices Here we define the alignment matrices based on Equation (4.6). Assume that the alignment or match as defined by (4.6) is calculated using the words to the right $w_n^+, n = 1, 2, \cdots, N$. Let $S[w^+]$ be the alignment matrix given by

$$
S[w^+] = \begin{pmatrix}
s_{11} & s_{12} & \cdots & s_{1N} \\
s_{21} & s_{22} & \cdots & s_{2N} \\
\vdots & \vdots & \ddots & \vdots \\
s_{M1} & s_{M2} & \cdots & s_{MN}
\end{pmatrix}.
\tag{4.7}
$$

The alignment of the object o_m with the words to the left can be defined by $1 - s_{mn}$ as noted above. From this we define another alignment matrix $S[w^-]$:

$$
S[w^-] = \begin{pmatrix}
1 - s_{11} & 1 - s_{12} & \cdots & 1 - s_{1N} \\
1 - s_{21} & 1 - s_{22} & \cdots & 1 - s_{2N} \\
\vdots & \vdots & \ddots & \vdots \\
1 - s_{M1} & 1 - s_{M2} & \cdots & 1 - s_{MN}
\end{pmatrix}.
\tag{4.8}
$$

Note 4.5 The method presented in this section cannot treat a word with a quantifier, for instance, "a little cute." To deal with such a case, we will consider fuzzy-set theoretical data processing later.

Question 4.1 Answer the following questions:[5]

(1) Show $-1 \le \rho_{xy} \le 1$.

(2) Derive (4.3) and (4.4).

(3) Show $\rho_{\tilde{x}\tilde{y}} = \rho_{xy}$.

(4) Derive the eigenvalue problem (4.5) from (4.3), (4.4).

(5) Show that the maximum eigenvalue of the problem (4.5) is $\lambda = 1$, which is a meaningless solution.

---

[5]The answers will be given in Section 4.5.

4.3.2   Probabilistic data processing

Using the frequency matrices defined in (4.2), we can calculate approximate probabilities by

$$p_{mnl} = \frac{y_{mnl}}{\displaystyle\sum_{l=-L}^{L} y_{mnl}}.$$

Then we can define the alignment matrices:

$$S_n = \begin{pmatrix} p_{1n(-L)} & p_{1n(-L+1)} & \cdots & p_{1nL} \\ p_{2n(-L)} & p_{2n(-L+1)} & \cdots & p_{2nL} \\ \vdots & \vdots & \ddots & \vdots \\ p_{Mn(-L)} & p_{Mn(-L+1)} & \cdots & p_{MnL} \end{pmatrix}$$

for $n = 1, 2, \cdots, N$. See Figure 4.12. But, practically, we should modify p_{mnl} taking into account the distribution of opinions. See, for instance, Huynh, Yan, and Nakamori (2010).

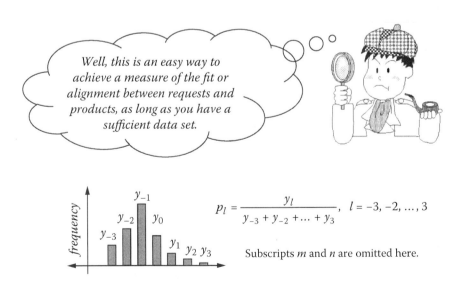

$$p_l = \frac{y_l}{y_{-3} + y_{-2} + \ldots + y_3}, \quad l = -3, -2, \ldots, 3$$

Subscripts m and n are omitted here.

Figure 4.12: An easy way to calculate alignment.

4.3.3   Fuzzy-set theoretical data processing

This section introduces a modeling method for alignment between objects and words using the fuzzy-sets theory (Zadeh, 1965). The superior feature of this method is that we can deal with possibilities in data, and calculate the alignment between objects and words with degrees (e.g., a little, very).

Fuzzy set   A fuzzy set is characterized by a membership function. Let X be the whole set to be considered. A fuzzy set A in X is defined by a membership function $\mu_A(x)$:

$$\mu_A : X \to [0,1].$$

That is, each element in X has a degree of membership to A, which is defined by a real number between 0 and 1. This definition includes the case that A is a crisp set:

$$\mu_A(x) = \begin{cases} 1, & x \in A; \\ 0, & x \notin A. \end{cases}$$

This book deals with the normal and convex fuzzy sets.

Normal fuzzy sets   A fuzzy set A is called *normal* if there exist $x \in X$ which satisfies $\mu_A(x) = 1$, or

$$\max_{x \in X} \mu_A(x) = 1.$$

Convex fuzzy sets   A fuzzy set A is convex if for any $\lambda \in [0,1]$ the following inequality holds:

$$\mu_A(\lambda x + (1 - \lambda)y) \geq \min\{\mu_A(x), \mu_A(y)\}, \quad x, y \in X.$$

Note that this definition is different from the ordinary convex function.

Membership functions of products   From (4.2) we calculate

$$\bar{y}_{mn} = \frac{\displaystyle\sum_{l=-L}^{L} (y_{mnl} \times l)}{\displaystyle\sum_{l=-L}^{L} y_{mnl}},$$

$$\sigma_{mn}^2 = \frac{\displaystyle\sum_{l=-L}^{L}\left\{y_{mnl}\times(l-\bar{y}_{mn})^2\right\}}{\displaystyle\sum_{l=-L}^{L}y_{mnl}}.$$

We need to define a membership function that represents the degree of alignment of the object o_m and the word $w_n : \langle w_n^-, w_n^+\rangle$ as follows:

$$\mu_{mn}(y) = \begin{cases} \dfrac{1}{c\sigma_{mn}}\left\{y-(\bar{y}_{mn}-c\sigma_{mn})\right\}, & y \le \bar{y}_{mn}; \\[3mm] -\dfrac{1}{c\sigma_{mn}}\left\{y-(\bar{y}_{mn}+c\sigma_{mn})\right\}, & y \ge \bar{y}_{mn}. \end{cases}$$

Here, $c\,(>1)$ is a tuning parameter. An example of this membership function is given in Figure 4.13, where we have set $L = 3$.

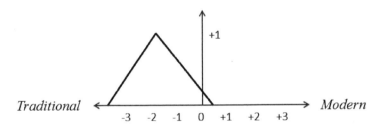

Figure 4.13: This product is rather traditional.

Membership functions of quantifiers Now we define membership functions $\{\mu_l(y) \mid l \in \{-L, \cdots, 0, \cdots, L\}\}$, each of which represents the l-level fuzzy set.

- For $l = -L$,

$$\mu_l(y) = \begin{cases} -y+1+l, & l \le y \le l+1; \\[2mm] 0, & \text{otherwise.} \end{cases}$$

- For $-L < l < L$,

$$\mu_l(y) = \begin{cases} y + 1 - l, & l - 1 \le y \le l; \\ -y + 1 + l, & l \le y \le l + 1; \\ 0, & \text{otherwise.} \end{cases}$$

- For $l = L$,

$$\mu_l(y) = \begin{cases} y + 1 - l, & l - 1 \le y \le l; \\ 0, & \text{otherwise.} \end{cases}$$

Figure 4.14 shows these membership functions when $L = 3$, where the levels of requirement between traditional and modern are shown clearly.

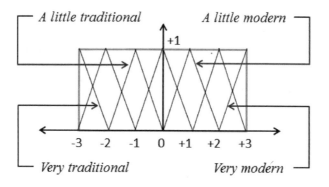

Figure 4.14: Seven-level membership functions.

Alignment   Now we can define the alignment or match between the object o_m and the word w_n with degrees $l \in \{-L, \cdots, 0, \cdots, L\}$:

$$s_{mnl} = \max_y \min \{\mu_{mn}(y), \mu_l(y)\}.$$

An example is given in Figure 4.15.

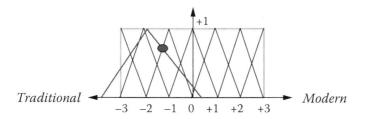

The height of ● *is the degree of alignment between*
this object and the value "a little traditional."

Figure 4.15: Calculation of fitness degrees.

Note 4.6 To avoid the situation where products do not meet the requirements, if the membership function $\mu_{mn}(y)$ is smaller than a given small positive number, then we replace its value with this number. Or, we can introduce
a Gaussian-type membership function, so for instance we can calculate the
degree of alignment between the object and *modern* as shown in Figure 4.16.

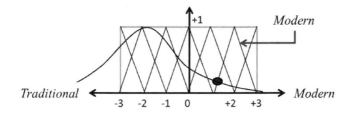

Figure 4.16: Use of a Gaussian-type membership function.

Alignment matrices We can now define the alignment matrices:

$$
S_n = \begin{pmatrix}
s_{1n(-L)} & s_{1n(-L+1)} & \cdots & s_{1nL} \\
s_{2n(-L)} & s_{2n(-L+1)} & \cdots & s_{2nL} \\
\vdots & \vdots & \ddots & \vdots \\
s_{Mn(-L)} & s_{Mn(-L+1)} & \cdots & s_{MnL}
\end{pmatrix}
\tag{4.9}
$$

for $n = 1, 2, \cdots, N$.

4.3.4   Direct modeling

Since the number of products in a store is usually very large, it is physically as well as financially difficult to make models based on an experiment using a large number of consumers. A more realistic way is that the shop owner makes his/her own models as shown in Figure 4.17, in which the owner encloses the range where he/she thinks that consumers are likely to answer by using an ellipse.

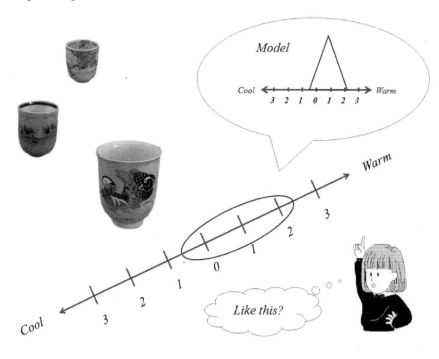

Figure 4.17: Direct fuzzy modeling by a shop owner.

Models can certainly be created in this way. Nevertheless, an experiment is still necessary in order to select the appropriate evaluation words as well as to know the general feelings of consumers.

4.4 Information Aggregation

Suppose that a customer's requirements are given by a set of words:

$$W = \{w'_1, w'_2, \cdots, w'_J\} \subset \{w_n^- \text{ or } w_n^+; n = 1, 2, \cdots, N\}$$

and suppose that we have the alignment matrices given in (4.7) and (4.8). See Figure 4.18.

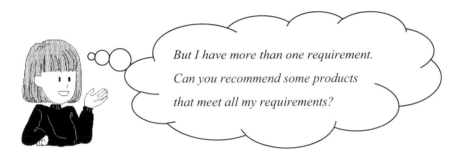

But I have more than one requirement.

Can you recommend some products

that meet all my requirements?

Figure 4.18: How do you aggregate multiple pieces of information?

Comprehensive evaluation An easy way to calculate the comprehensive evaluation or the total score is given by the weighted sum:

$$CE(m) = \mu_1 s_{m1} + \mu_2 s_{m2} + \cdots + \mu_J s_{mJ}.$$

Here, s_{mj} is the alignment or match given in (4.7) or (4.8), and μ_j is the weighting for the word w'_j. However, in most cases, the customer's multiple requirements are alternative or complementary, therefore we will consider non-linear aggregation methods in this section. They are:

1. Choquet integral[6] with non-additive measure (Choquet, 1954)

2. Ordered weighted averaging aggregation (Yager, 1988)

3. Prioritized max–min aggregation (Nakamori, 2011)

---

[6]The Choquet integral has been intensively studied in the fuzzy set societies around the world. See for instance, Grabisch (1996, 2003).

4.4.1 Non-additive measures

Consider the example where a customer's requirements are multiple and interrelated as shown in Figure 4.19.

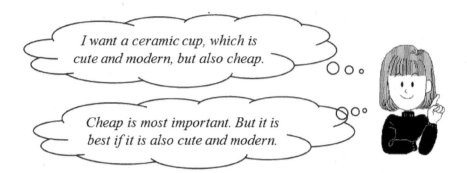

I want a ceramic cup, which is cute and modern, but also cheap.

Cheap is most important. But it is best if it is also cute and modern.

Figure 4.19: How do you resolve such contradictions?

Requirements Let us confirm the situation:

- The customer's requirements are given by a set of words:

$$W = \{w_1', w_2', \cdots, w_J'\} \subset \{w_n^- \text{ or } w_n^+; n = 1, 2, \cdots, N\}.$$

- The customer expresses complex requirements as shown in Figure 4.19.

Measure Mathematically, the measure is a number given to an object. Examples of the measure are length, area, temperature, etc. Let X be a set, and 2^X be the power set of X, i.e., the set of all subsets of X. The measure is a function to assign a non-negative real number to each element of 2^X. The measure satisfies:

1. $\mu(\phi) = 0$; $\mu(A) \geq 0$, $^\forall A \in 2^X$.

2. $\mu(X) = $ a finite value.

3. $A, B \in 2^X$, $A \cap B = \phi \rightarrow \mu(A \cup B) = \mu(A) + \mu(B)$.

The most important property is the last one, which is called *additivity*.

Non-additive measure   The non-additive measure, often called the fuzzy measure, is defined by

1. $\mu(\phi) = 0$; $\mu(A) \geq 0$, $^\forall A \in 2^X$.

2. $\mu(X) = $ a finite value.

3. $A, B \in 2^X$, $A \subset B \rightarrow \mu(A) \leq \mu(B)$.

Here, the additivity is replaced with the *monotonicity.*

Choquet integral   Sort:

$$\{1, 2, \cdots, J\} \rightarrow (n_1, n_2, \cdots, n_J)$$

such that the following holds

$$s_{mn_1} \geq s_{mn_2} \geq \cdots \geq s_{mn_J}.$$

The comprehensive evaluation of o_m is given by the Choquet integral:[7]

$$
\begin{aligned}
CE(m) \;\; = \;\; & \mu(\{n_1\})(s_{mn_1} - s_{mn_2}) + \mu(\{n_1, n_2\})(s_{mn_2} - s_{mn_3}) \\
& + \cdots + \mu(\{n_1, n_2, \cdots, n_J\}) s_{mn_J}.
\end{aligned}
$$

Note 4.7   If the measure is additive, the above equation becomes

$$CE(m) = \mu(\{n_1\}) s_{mn_1} + \mu(\{n_2\}) s_{mn_2} + \cdots + \mu(\{n_J\}) s_{mn_J}.$$

Example 4.1   Suppose that we have evaluations for the products o_1 and o_2 as shown in Table 4.1.

Table 4.1: An example of evaluation of two objects.

	Cheap (w_1')	Cute (w_2')	Modern (w_3')
o_1	$s_{11} = 0.6$	$s_{12} = 0.8$	$s_{13} = 0.7$
o_2	$s_{21} = 0.6$	$s_{22} = 0.5$	$s_{23} = 1.0$

---

[7]Here we represent $\mu(\{n_1\})$ instead of $\mu(\{w_{n_1}'\})$ for simplicity.

Non-additive measures   From Table 4.1 we see that

$$s_{12} > s_{13} > s_{11}; \quad s_{23} > s_{21} > s_{22}.$$

If we use the simple average, that is, the weights of three words are set equally at $1/3$, then the overall assessment values of two objects are the same as 0.7. If we consider that cheapness is most important and set the weights of cheap, cute, and modern as 0.5, 0.25, and 0.25, respectively, then the overall assessment values of two objects are again the same as 0.675. Thus, we cannot give an order to those objects in either case. Therefore, we introduce the non-additive measures as follows:

- $\mu(\phi) = 0.0$

- $\mu(\{1\}) = 0.5, \quad \mu(\{2\}) = 0.25, \quad \mu(\{3\}) = 0.25$

- $\mu(\{1,2\}) = 0.5, \quad \mu(\{1,3\}) = 0.5, \quad \mu(\{2,3\}) = 1.0$

- $\mu(\{1,2,3\}) = 1.0$

Here, cheapness is most important if we consider a single item of evaluation. However, if the object is cute and at the same time modern, we give a large weight to that object.

Choquet integrals   Using these measures we can calculate $CE(1)$ and $CE(2)$ as follows:

$$
\begin{aligned}
CE(1) &= \mu(\{2\})\,(s_{12} - s_{13}) + \mu(\{2,3\})\,(s_{13} - s_{11}) + \mu(\{1,2,3\})s_{11} \\
&= 0.725,
\end{aligned}
$$

$$
\begin{aligned}
CE(2) &= \mu(\{3\})\,(s_{23} - s_{21}) + \mu(\{1,3\})\,(s_{21} - s_{22}) + \mu(\{1,2,3\})s_{22} \\
&= 0.650.
\end{aligned}
$$

From these we see that $CE(1) > CE(2)$, that is, the product o_1 is better than the product o_2.

Question 4.2   Consider the reason why $CE(1) > CE(2)$.

4.4.2 Ordered Weighted Averaging (OWA) operators

Here we also use the matrices defined in (4.7) and (4.8). However, we must deal with the situation shown in Figure 4.20, where linguistic quantifiers are used in expressing the desired attributes.

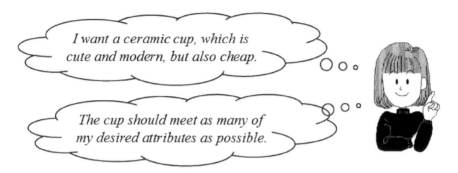

I want a ceramic cup, which is cute and modern, but also cheap.

The cup should meet as many of my desired attributes as possible.

Figure 4.20: How could you help her?

Linguistic quantifier   Here we will deal with the following linguistic quantifiers (LQ) (Zadeh, 1983):

- LQ = all

- LQ = as many as possible

- LQ = most

- LQ = at least half

- LQ = there exists

Examples of usage of these quantifiers are:

- I will be happy if most of my requests are satisfied.

- I will be pleased even if only one of my requests is satisfied.

In this section we will use ordered weighted averaging operators, often called OWA operators, which provide a parameterized class of mean type aggregation operators. See, for instance, Yager and Kacprzyk (1997).

The membership functions of the above linguistic quantifiers should satisfy these criteria:

1. $f : [0, 1] \rightarrow [0, 1]$

2. $f(0) = 0$

3. $f(x) = 1,\ \exists x \in [0, 1]$

4. $f :$ non-decreasing

For instance, let $a < b,\ a, b \in [0, 1]$,

$$f(x) = \begin{cases} 0, & \text{if } x \leq a; \\ \dfrac{x - a}{b - a}, & \text{if } a \leq x \leq b; \\ 1, & \text{if } x \geq b. \end{cases}$$

Table 4.2 shows some examples of such membership functions.

Table 4.2: Linguistic quantifiers and their membership functions.

Linguistic Quantifier	Membership Function
LQ = all	$f(x) = \begin{cases} 0, & \text{if } x \neq 1 \\ 1, & \text{if } x = 1 \end{cases}$
LQ = as many as possible	$f(x) = \begin{cases} 0, & \text{if } x \leq 0.5 \\ 2x - 1, & \text{if } x \geq 0.5 \end{cases}$
LQ = most	$f(x) = \begin{cases} 0, & \text{if } x \leq 0.3 \\ 2x - 0.6, & \text{if } 0.3 \leq x \leq 0.8 \\ 1, & \text{if } x \geq 0.8 \end{cases}$
LQ = at least half	$f(x) = \begin{cases} 2x, & \text{if } x \leq 0.5 \\ 1, & \text{if } x \geq 0.5 \end{cases}$
LQ = there exists	$f(x) = \begin{cases} 0, & \text{if } x < 1/J \\ 1, & \text{if } x \geq 1/J \end{cases}$

Aggregating weights For any membership function given in Table 4.2, the weightings are calculated by

$$\mu_j = f\left(\frac{j}{J}\right) - f\left(\frac{j-1}{J}\right), \quad j = 1, 2, \cdots, J.$$

Sort

$$\{1, 2, \cdots, J\} \rightarrow (n_1, n_2, \cdots, n_J)$$

such that the following holds

$$s_{mn_1} \geq s_{mn_2} \geq \cdots \geq s_{mn_J}.$$

Then the comprehensive evaluation value of the object o_m is given by

$$CE(m) = \sum_{j=1}^{J} \mu_j s_{mn_j}.$$

The most recommended product o_{m^*} is given by

$$m^* = \arg\max_m \{CE(m)\}.$$

OWA operator Generally, the n-dimensional OWA operator F is a mapping

$$F : [0, 1]^n \rightarrow [0, 1].$$

This mapping is represented by

$$F(a_1, a_2, \cdots, a_n) = \sum_{j=1}^{n} \mu_j\, b_j.$$

Here, b_j is the j-th largest value among $a_1, a_2, \cdots, a_n$, and

$$\mu_j \in [0, 1], \quad \sum_{j=1}^{n} \mu_j = 1.$$

Using this we have

$$E_{mj(l)} = g_p\left(s_{mj(l)(p)}\right).$$

The comprehensive evaluation of the product o_m is then given by

$$CE(m) = \min_{j(l)}\left\{E_{mj(l)}\right\}.$$

Finally, the most recommended product o_{m^*} is given by

$$m^* = \arg\max_{m}\left\{CE(m)\right\}.$$

Note that this max–min strategy corresponds to *for all* in OWA operation.

Example 4.3 Let us consider an example to compare two products again, but this time the attributes have levels and priorities as shown in Table 4.3.

Table 4.3: An example of evaluation of two objects.

	Cheap (w_1')	A little cute (w_2')	Quite modern (w_3')
	($l = -2$) ($p = 1$)	($l = 1$) ($p = 2$)	($l = 3$) ($p = 3$)
o_1	$s_{11(-2)(1)} = 0.5$	$s_{12(1)(2)} = 0.9$	$s_{13(3)(3)} = 0.7$
o_2	$s_{21(-2)(1)} = 0.5$	$s_{22(1)(2)} = 0.6$	$s_{23(3)(3)} = 1.0$

- Here, we assign the levels of requirement:

  Cheap $\rightarrow l = -2$

  A little cute $\rightarrow l = 1$

  Quite modern $\rightarrow l = 3$

- We assume that the customer declares the priority as follows:

  Cheap $\rightarrow p = 1$

  A little cute $\rightarrow p = 2$

  Quite modern $\rightarrow p = 3$ (highest)

- Figure 4.23 shows an example of how to calculate the alignment.

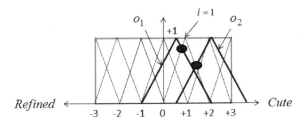

Figure 4.23: An example to calculate the alignment.

Note that $p = 3$ means *most important* in this case. Then we have the transformation functions:

$$g_1(x) = \begin{cases} 5x, & 0 \le x \le \frac{1}{6}; \\ \frac{1}{5}x + \frac{4}{5}, & \frac{1}{6} \le x \le 1; \end{cases}$$

$$g_2(x) = \begin{cases} 2x, & 0 \le x \le \frac{1}{3}; \\ \frac{1}{2}x + \frac{1}{2}, & \frac{1}{3} \le x \le 1; \end{cases}$$

$$g_3(x) = x, \quad 0 \le x \le 1.$$

From the above, we have

$$CE(1) = E_{13(3)} = 0.7, \quad CE(2) = E_{22(1)} = 0.8,$$

$$m^* = \arg \max_m \{CE(m); m = 1, 2\} = 2.$$

Thus, it is appropriate to recommend product o_2.

Question 4.4 Consider why the product o_2 is selected in this example.

If you want products that meet all of your needs, the word with the lowest evaluation value is important in the max–min strategy. If the priority of this word is low, its evaluation value should be increased so that this word will not affect the decision-making process.

4.5   Answers to Questions

Answer to Question 4.1 (1)   Let

$$\alpha_m = x_m - \sum_{m=1}^{M} p_{m\bullet} x_m, \quad \beta_n = y_n - \sum_{n=1}^{N} p_{\bullet n} y_n.$$

Then we have

$$
\begin{aligned}
\sigma_x^2 &= \sum_{m=1}^{M} p_{m\bullet} x_m^2 - \left(\sum_{m=1}^{M} p_{m\bullet} x_m \right)^2 \\
&= \sum_{m=1}^{M} p_{m\bullet} \left(x_m - \sum_{m=1}^{M} p_{m\bullet} x_m \right)^2 \\
&= \sum_{m=1}^{M} p_{m\bullet} \alpha_m^2 = \sum_{m=1}^{M} \sum_{n=1}^{N} p_{mn} \alpha_m^2,
\end{aligned}
$$

$$
\begin{aligned}
\sigma_y^2 &= \sum_{n=1}^{N} p_{\bullet n} y_n^2 - \left(\sum_{n=1}^{N} p_{\bullet n} y_n \right)^2 \\
&= \sum_{n=1}^{N} p_{\bullet n} \left(y_n - \sum_{n=1}^{N} p_{\bullet n} y_n \right)^2 \\
&= \sum_{n=1}^{N} p_{\bullet n} \beta_n^2 = \sum_{m=1}^{M} \sum_{n=1}^{N} p_{mn} \beta_n^2,
\end{aligned}
$$

$$
\begin{aligned}
\sigma_{xy} &= \sum_{m=1}^{M} \sum_{n=1}^{N} p_{mn} x_m y_n - \left(\sum_{m=1}^{M} p_{m\bullet} x_m \right) \left(\sum_{n=1}^{N} p_{\bullet n} y_n \right) \\
&= \sum_{m=1}^{M} \sum_{n=1}^{N} p_{mn} \left(x_m - \sum_{m=1}^{M} p_{m\bullet} x_m \right) \left(y_n - \sum_{n=1}^{N} p_{\bullet n} y_n \right) \\
&= \sum_{m=1}^{M} \sum_{n=1}^{N} p_{mn} \alpha_m \beta_n.
\end{aligned}
$$

Let s and t be any real numbers. The following holds:

$$\sum_{m=1}^{M} \sum_{n=1}^{N} p_{mn} (s\alpha_m + t\beta_n)^2 \geq 0.$$

That is,

$$s^2 \sigma_x^2 + 2st\sigma_{xy} + t^2 \sigma_y^2 \geq 0.$$

Letting

$$s = \sigma_y^2, \quad t = -\sigma_{xy},$$

we have

$$\sigma_x^2 \sigma_y^4 - \sigma_{xy}^2 \sigma_y^2 \geq 0.$$

Here we assume that $\sigma_x \neq 0$, $\sigma_y \neq 0$, so we have

$$\rho_{xy}^2 = \frac{\sigma_{xy}^2}{\sigma_x^2 \sigma_y^2} \leq 1.$$

Answer to Question 4.1 (2)

$$
\begin{aligned}
\frac{\partial \rho_{xy}}{\partial x_m} &= \frac{1}{\sigma_x \sigma_y} \frac{\partial \sigma_{xy}}{\partial x_m} - \frac{\sigma_{xy}}{\sigma_x^2 \sigma_y} \frac{\partial \sigma_x}{\partial x_m} \\
&= \frac{1}{\sigma_x \sigma_y} \left(\sum_{n=1}^{N} p_{mn} y_n - p_{m\bullet} \sum_{n=1}^{N} p_{\bullet n} y_n \right) \\
&\quad - \frac{\rho_{xy}}{\sigma_x^2} p_{m\bullet} \left(x_m - \sum_{m=1}^{M} p_{m\bullet} x_m \right) \\
&= \frac{1}{\sigma_x} \left(\sum_{n=1}^{N} p_{mn} \tilde{y}_n - \rho_{xy} p_{m\bullet} \tilde{x}_m \right).
\end{aligned}
$$

$$
\begin{aligned}
\frac{\partial \rho_{xy}}{\partial y_n} &= \frac{1}{\sigma_x \sigma_y} \frac{\partial \sigma_{xy}}{\partial y_n} - \frac{\sigma_{xy}}{\sigma_x \sigma_y^2} \frac{\partial \sigma_y}{\partial y_n} \\
&= \frac{1}{\sigma_x \sigma_y} \left(\sum_{m=1}^{M} p_{mn} x_m - p_{\bullet n} \sum_{m=1}^{M} p_{m\bullet} x_m \right) \\
&\quad - \frac{\rho_{xy}}{\sigma_y^2} p_{\bullet n} \left(y_n - \sum_{n=1}^{N} p_{\bullet n} y_n \right) \\
&= \frac{1}{\sigma_y} \left(\sum_{m=1}^{M} p_{mn} \tilde{x}_m - \rho_{xy} p_{\bullet n} \tilde{y}_n \right).
\end{aligned}
$$

Answer to Question 4.1 (3)   From the definition,

$$\tilde{x}_m = \frac{1}{\sigma_x}\left(x_m - \sum_{m=1}^{M} p_{m\bullet} x_m\right), \quad \tilde{y}_n = \frac{1}{\sigma_y}\left(y_n - \sum_{n=1}^{N} p_{\bullet n} y_n\right).$$

Then

$$\sigma_{\tilde{x}}^2 = \sigma_{\tilde{y}}^2 = 1, \quad \sigma_{\tilde{x}\tilde{y}} = \frac{\sigma_{xy}}{\sigma_x \sigma_y},$$

$$\rho_{\tilde{x}\tilde{y}} = \frac{\sigma_{\tilde{x}\tilde{y}}}{\sigma_{\tilde{x}}\sigma_{\tilde{y}}} = \frac{\sigma_{xy}}{\sigma_x \sigma_y} = \rho_{xy}.$$

Answer to Question 4.1 (4)   Multiplying P_y^{-1} by (4.4) from the left, we have

$$P_y^{-1}P^t\tilde{\boldsymbol{x}} - \rho_{xy}\tilde{\boldsymbol{y}} = \boldsymbol{0}.$$

Substituting $\tilde{\boldsymbol{y}}$ in (4.3), we have

$$\left(PP_y^{-1}P^t - \rho_{xy}^2 P_x\right)\tilde{\boldsymbol{x}} = \boldsymbol{0}.$$

By deforming this, we have

$$\left(P_x^{-\frac{1}{2}}PP_y^{-1}P^tP_x^{-\frac{1}{2}} - \rho_{xy}^2 I\right)P_x^{\frac{1}{2}}\tilde{\boldsymbol{x}} = \boldsymbol{0}.$$

Answer to Question 4.1 (5)   Representing (4.3), (4.4) by elements, we have

$$\sum_{n=1}^{N} p_{mn}\tilde{y}_n - \rho_{xy}p_{m\bullet}\tilde{x}_m = 0, \quad m = 1, 2, \cdots, M,$$

$$\sum_{m=1}^{M} p_{mn}\tilde{x}_m - \rho_{xy}p_{\bullet n}\tilde{y}_n = 0, \quad n = 1, 2, \cdots, N.$$

Here,

$$\left\{\rho_{xy} = 1; \ \tilde{x}_m = 1, \ ^\forall m; \ \tilde{y}_n = 1, \ ^\forall n\right\}$$

satisfies the above two equations, and gives us a solution to the eigenvalue problem. From

$$\lambda = \rho_{xy}^2 \leq 1,$$

$\lambda = 1$ is the maximum solution; however, it is a meaningless solution.

Answer to Question 4.2   Figure 4.24 shows graphical explanations of the Choquet integral for products o_1 and o_2, respectively. From Figure 4.24, you can easily understand why the product o_1 is better.

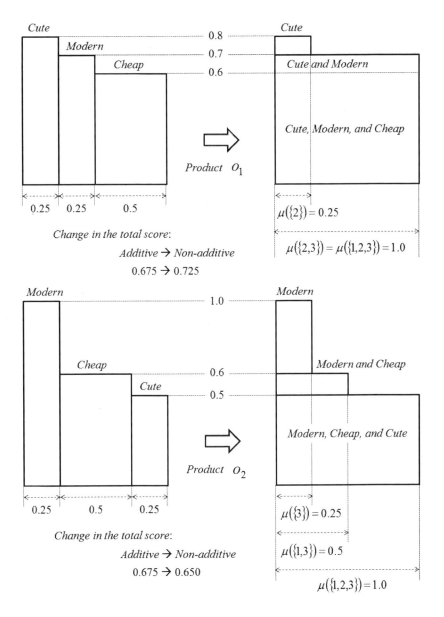

Figure 4.24: Choquet integral for products o_1 and o_2.

Answer to Question 4.3

1. LQ = all: $\mu_1 = 0.0,\quad \mu_2 = 0.0,\quad \mu_3 = 1.0$

$$CE(1) = 0.6,\quad CE(2) = 0.5\quad \rightarrow\quad CE(1) > CE(2)$$

2. LQ = most: $\mu_1 = 0.067,\quad \mu_2 = 0.667,\quad \mu_3 = 0.267$

$$CE(1) = 0.681,\quad CE(2) = 0.601\quad \rightarrow\quad CE(1) > CE(2)$$

3. LQ = at least half: $\mu_1 = 0.667,\quad \mu_2 = 0.333,\quad \mu_3 = 0.0$

$$CE(1) = 0.767,\quad CE(2) = 0.867\quad \rightarrow\quad CE(1) < CE(2)$$

4. LQ = there exists: $\mu_1 = 1.0,\quad \mu_2 = 0.0,\quad \mu_3 = 0.0$

$$CE(1) = 0.8,\quad CE(2) = 1.0\quad \rightarrow\quad CE(1) < CE(2)$$

Answer to Question 4.4 Figure 4.25 shows the transformations. Because the priority of the requirement *cheap* is the lowest, its evaluation values are transformed to larger values.[8] As a result, *cheap* is no longer involved in the decision.

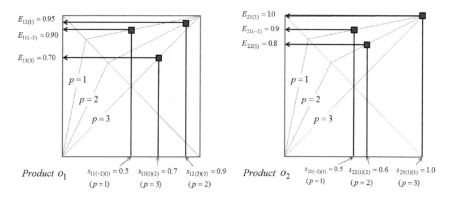

Figure 4.25: Transformations according to priorities of desires.

---

[8]This idea was suggested by Marek Makowski, International Institute for Applied Systems Analysis, Austria.

4.6 Application of Specialized Integration

This section introduces a government project[9] to develop a recommendation system to provide products to customers by recognizing their kansei desires. It aims to support sales expansion and new product development in traditional crafts in Ishikawa Prefecture, Japan. In order to do this, the project is developing a technique for selecting and providing information according to an individual person's kansei desires. This will be done by creating a kansei search engine and an information aggregation system. See Figure 4.26.

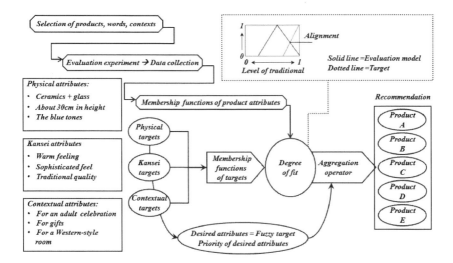

Figure 4.26: Functions of the recommendation system.

This recommendation system has already been installed on several websites of arts and crafts shops. If a user inputs his/her desired attributes, then the system will recommend several products, but the system prepares the bipolar scales in advance, which are different for respective stores. To deal with a large number of products, the present system uses the direct modeling approach, and it selectively uses the ordered weighted averaging operators and the prioritized max–min operators to aggregate information.

---

[9]This study was supported by SCOPE 102305001 of the Ministry of Internal Affairs and Communications, Japan.

Knowledge integration   The method described above gives a partial answer to the subject of product recommendation by knowledge integration. Actually, this corresponds to *Intelligence* in Figure 4.27, which shows an example of specified knowledge integration.

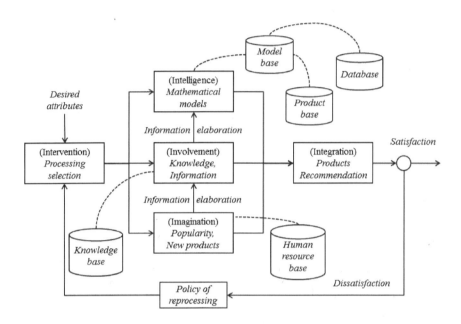

Figure 4.27: A framework of knowledge integration.

- *Intervention*: The user inputs his/her desired attributes, and if necessary selects a method of information aggregation.

- *Intelligence*: The computer system recommends several products that might fit with the desired attributes, based on the preinstalled aggregation methods and product information.

- *Involvement*: We can use the idea of widely deployed recommendation systems; examples given in Wikipedia (2013d) are:

  - "When viewing a product on Amazon.com, the store will recommend additional items based on a matrix of what other shoppers bought along with the currently selected item."

- "Pandora Radio takes an initial input of a song or musician and plays music with similar characteristics (based on a series of keywords attributed to the inputted artist or piece of music). The genre stations created by Pandora can be refined through user feedback (emphasizing or de-emphasizing certain characteristics)."

- "Netflix offers predictions of movies that a user might like to watch based on the user's previous ratings and watching habits (as compared to the behavior of other users), also taking into account the characteristics of the film (such as the genre)."

- *Imagination*: The information used in *Intelligence* and *Involvement* are based on past data. However, producers have ideas about current and future fashion trends, and also know their products. We can use such knowledge in *Imagination*.

- *Integration*: If we could install all the necessary information in a computer, we could recommend some products by using a certain integration rule. But usually the integrator, the shop owner, might have information from *Imagination* as a result of direct communication with the producers. So, we could invent an integration rule by taking into account information that might be regarded as tacit knowledge.

Searching traditional crafts with words.

*Systemic synthesis is achieved by the
interaction of explicit knowledge and tacit
knowledge, and knowledge in between, so
knowledge coordinators are needed.*

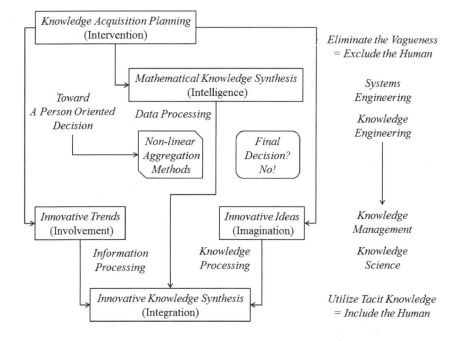

Chapter 5

The Emergence of Knowledge Science

5.1 Knowledge Revolution

A major social change of the early twenty-first century has been called the *knowledge revolution*. A new discipline, *knowledge science*, has attracted attention as one of the driving forces of this new society. There is also a hope that we can dispel the sense of stagnation in the integration of knowledge in *systems science* by developing this new discipline. However, see Figure 5.1.

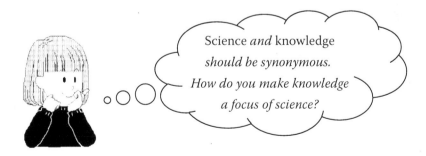

Figure 5.1: How do you make knowledge a focus of science?

The second half of this book is devoted to the fusion of systems science and knowledge science in order to achieve systemic integration of knowledge.

5.1.1 The value of knowledge

We experienced an information revolution in the twentiethth century, which brought about big changes in the economy and society with the advancement of information technology and communication. Industries have been developing information technology to improve efficiency and speed. In parallel with this, the rise of a knowledge-based industry has grown since the 1960s, in fields such as education, telecommunications, publishing, and research and development. Along with these trends, the development of the global network has led the knowledge revolution of the twenty-first century.

> Knowledge has become the main engine of the economy and society.

Question 5.1   How should we treat knowledge as an economic product, which has different characteristics than the traditional factors of production such as labor, capital, or land? See Figure 5.2.

Figure 5.2: Can you borrow money from the bank with your ideas?

Hint   What are the differences between knowledge and other production factors such as labor, capital, or land? Can you handle knowledge separately from those factors?

5.1.2 Information and knowledge

Now let us consider the difference between information and knowledge.

- By observing some events or phenomena, we collect data that is represented by letters and numbers as a component of information.

- By structuring the collected data, we obtain information as a set of data including the means of construction, where accuracy and objectivity are emphasized. If we could prove this information we might say we have formulated a rule.

- On the other hand, by experiencing some events or phenomena, we obtain knowledge that is understanding and beliefs about the nature of things and events. Knowledge is the ability to take action on the basis of the obtained information. We might develop a belief based on thoughts in regard to this knowledge.

Thus, knowledge is a different concept from information. See Figure 5.3.

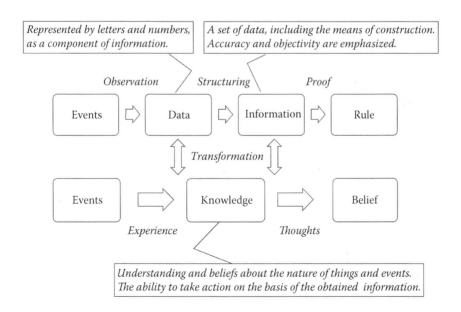

Figure 5.3: Knowledge is a different concept from information.

5.1.3   Knowledge management

Knowledge management is the prime term in knowledge science. This technical term was introduced by the computer industry in the 1980s as a computer software technique. The term was adopted in management science in the early 1990s and achieved great success as an idea generation technique. As a result, there are two interpretations of this term:

- Management of information related to knowledge-intensive activities

- Management of people in knowledge creation processes

Information technology emphasizes the former, while the latter is emphasized in organizational theory and study, as well as systematic knowledge creation processes. However, we consider that these two interpretations are not sufficient to explain knowledge management. The third fundamental interpretation is:

- Management of human resources in the knowledge-based society

Herein lies the important task of training people generally known as *knowledge workers* and guaranteeing their numbers and quality.

Question 5.2   Why are knowledge management activities sometimes considered the same as information management activities?

Hint   Knowledge management is considered to be document-driven knowledge sharing in a narrow sense, using information technology to treat explicit knowledge.

- Wilson (2002) argues, "Knowledge management appears to be very much equated with information management, and some people assume wrongly that tacit knowledge can be made explicit."

Sources   There are already many books on knowledge management; among them, Ackerman, Pipek, and Wulf (2003), Nonaka (2005), and Liebowitz (2012) are all edited books that include important research papers.

5.1.4 Tacit and explicit knowledge

We will now clarify the difference between tacit knowledge and explicit knowledge.

- Tacit knowledge is subjective and difficult to verbalize; related to a specified time and place.

- Explicit knowledge is objective and can be verbalized; based on scientific knowledge.

See Table 5.1.

Table 5.1: Tacit knowledge and explicit knowledge.

Tacit Knowledge	*Explicit Knowledge*
Knowledge derived from direct sensory experience.	*Systematic knowledge segmented from tacit knowledge.*
Often limited to a particular person or place, or a specific subject.	*Migration, dissemination, and reuse are possible, using information systems.*
Sharing and development are possible through collaboration with direct experience.	*Sharing and editing are possible through the medium of language.*

Question 5.3   Consider examples of tacit knowledge. See Figure 5.4.

Can you explain how to ride a bicycle?

Can you explain the delicate blending of flavors?

Figure 5.4: Examples of tacit knowledge.

5.1.5   SECI spiral

Organizational knowledge creation is for the production of new knowledge, at each level of the entire organization, based upon the information from the environment. A famous model called the SECI spiral (Nonaka and Takeuchi, 1995) [1] is shown in Figure 5.5, which will be explained later in Chapter 6. This model stresses that new knowledge could be created by the interaction between tacit and explicit knowledge, through the process of *Socialization, Externalization, Combination,* and *Internalization* (or SECI).

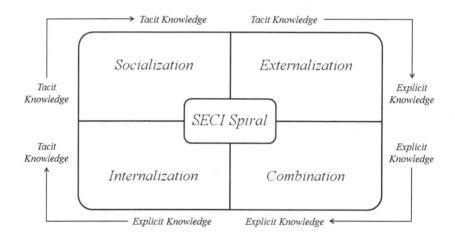

Figure 5.5: An organizational knowledge creation model.

- *Socialization* is the sharing of tacit knowledge through direct experience.

- *Externalization* is the creation of concepts through thoughtful dialogue.

- *Combination* is the creation of new knowledge through the combination of explicit knowledge.

- *Internalization* is the learning of new tacit knowledge, through implementation of explicit knowledge.

---

[1] Ikujiro Nonaka and others have since published many additional books; for instance, see Krogh, Ichijo and Nonaka (2000), Dierkes, Child, Antal, and Nonaka (2003), Takeuchi and Nonaka (2004), or Nonaka and Zhu (2012).

5.2   Epistemes in the Knowledge-Based Society

We are living in an information revolution that is leading to a new era. The most important change in this revolution might be a changing episteme: the way of constructing and justifying knowledge. The industrial episteme has been destroyed by Einstein's relativism (1916), Heisenberg's uncertainty principle (1927), complex theories, and finally the emergence[2] principle. See Figure 5.6.

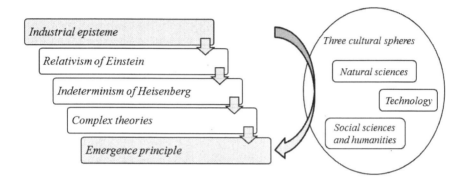

Figure 5.6: What is the episteme in the knowledge-based society?

The episteme of the industrial civilization was subjected to a process of destruction, particularly visible in the last fifty years. This has led to the divergent development of separate epistemes in three cultural spheres: the natural sciences, technology, and social sciences and humanities.

Question 5.4   What is the industrial episteme?

Hint   *Reductionism*: The belief that all systems, no matter how complex, can be understood by examining their parts in sufficient detail. This is, however, valid only if the level of complexity of the system is rather low.

---

[2]Emergence is the way complex systems and patterns arise out of a multiplicity of relatively simple interactions. Emergence is central to the theories of integrative levels and of complex systems.

5.2.1   Toward establishing a new episteme

The episteme of the knowledge civilization has not formed yet, but it must include integration of the following:

- A synthesis of the divergent epistemes of different cultural spheres

- A synthesis of the different aspects of Oriental and Occidental epistemes

The integration must be based on a holistic understanding of human nature: An attempt at such integration has been made at the Japan Advanced Institute of Science and Technology (JAIST),[3] in the School of Knowledge Science.

5.2.2   School of Knowledge Science

The School of Knowledge Science is the first school established in the world to make knowledge a focus of science. It consists of three disciplines as shown in Figure 5.7.

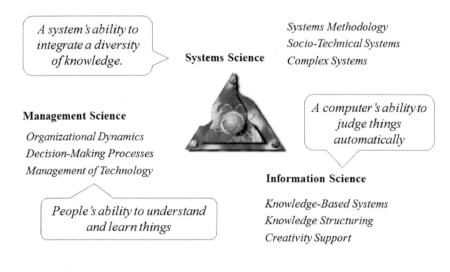

Figure 5.7: Structure of the School of Knowledge Science.

---

[3]JAIST was founded on October 1, 1990 as the first national institute in Japan that consists of graduate schools without undergraduate programs, and that possesses its own campus and organization for research and education.

The present stage At the present stage, knowledge science is more a problem-oriented interdisciplinary academic field than a single discipline.

> *The mission of knowledge science is to organize and process both objective and subjective information, and to create new value and new knowledge.*

Knowledge science mainly deals with the research area involving social innovations such as the regeneration of organizations, systems, and the mind. However, society's progress is underpinned by technology, and the joint progress of society (needs) and technology (seeds) is essential, so an additional mission for knowledge science is to act as a coordinator for extensive technological and social innovations.

- In order to fulfill these missions, the School of Knowledge Science focuses its research and education on observing and modeling the actual process of carrying out the mission as well as developing methods to carry out the mission.

The methods are mainly being developed through the existing three areas in the school which involve the application of the following:

- business science/organizational theories (practical use of tacit knowledge, management of technology, innovation theory),

- information technology/artistic methods (knowledge discovery methods, ways to support knowledge creation, knowledge engineering, cognitive science), and

- mathematical systems theory (systems thinking, the emergence principle, sociotechnical systems).

Question 5.5 How have you been carrying out knowledge management? Cite examples from your research (or study) at universities, or from doing business.

Question 5.6 How would you describe knowledge science in your understanding, at this point? Note that there is no guarantee that the present configuration of the School of Knowledge Science is appropriate.

5.3   Approaches to Knowledge Science

Now let us consider approaches to knowledge science.

1. Approaches from information science

   - Knowledge engineering
   - Knowledge discovery (data mining)
   - Knowledge construction (agent-based simulation)
   - Knowledge management (information management)

2. Approaches from management science

   - Organizational knowledge creation models
   - Academic knowledge creation models

3. An approach from systems science

   - Systemic intervention and synthesis

Question 5.7   What are the features of the above approaches?

Hint   (corresponding to the numbers above)

1. Use of computers, but expanding on traditional information science because information science uses subjective knowledge such as

   - experts' knowledge,
   - domain knowledge, and
   - tacit knowledge.

2. Use of actual people, beyond information science:

   - Individual tacit knowledge (experience)
   - Distributed tacit knowledge (consideration)
   - Individual intuition (enlightenment)

3. Use of a variety of knowledge and media: this is knowledge science.

5.3.1 Approaches from information science

Knowledge engineering   This involves integrating knowledge in computer systems to solve complex problems normally requiring a high level of human expertise (Feigenbaum and McCorduck, 1983). See Figure 5.8.

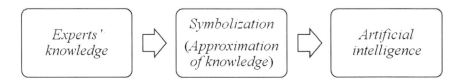

Figure 5.8: Knowledge engineering.

It refers to the building, maintaining, and development of knowledge-based systems. It has a great deal in common with software engineering, and is used in many computer science domains such as artificial intelligence including databases, data mining, expert systems, and decision support systems.

Knowledge discovery (data mining)   Data mining is the process of extracting patterns from large data sets by combining methods from statistics and artificial intelligence with database management. See Figure 5.9.

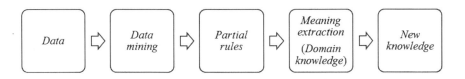

Figure 5.9: Knowledge discovery and data mining.

Data mining is seen as an increasingly important tool by modern business to transform data into business intelligence, therefore providing an informational advantage. It is currently used in a wide range of profiling practices, such as marketing, surveillance, fraud detection, and scientific discovery.

Knowledge construction (agent-based modeling and simulation)   An agent-based model is a class of computational models for simulating the actions and interactions of autonomous agents with a view to assessing their effects on the system as a whole. The models simulate the simultaneous operations and interactions of multiple agents, in an attempt to re-create and predict the appearance of complex phenomena. See Figure 5.10.

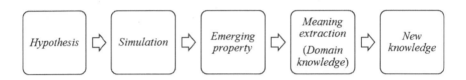

Figure 5.10: Knowledge construction (agent-based simulation).

The process is the emergence from the lower level of systems to a higher level. A key notion is that simple behavioral rules generate complex behavior. This principle, known as KISS ("Keep it simple, stupid.") is extensively adopted in the modeling community.

Knowledge management (information management)   Knowledge management comprises a range of strategies and practices used in an organization to identify, create, represent, distribute, and enable adoption of insights and experiences. Such insights and experiences comprise knowledge, either embodied in individuals or embedded in organizational processes or practices. See Figure 5.11.

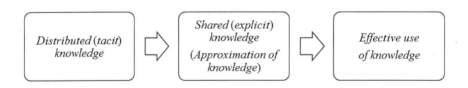

Figure 5.11: Knowledge management (information management).

Knowledge management focuses on organizational objectives such as improved performance, competitive advantage, innovation, sharing of lessons learned, integration, and continuous improvement of the organization.

5.3.2 Approaches from management science

Organizational knowledge creation (bottom-up model) An essential element of *The Knowledge Creating Company* (Nonaka and Takeuchi, 1995) consists of motivating a scientific revolution in knowledge creation theories by presenting a process- and algorithmic-like principle of organizational knowledge creation. See Figure 5.12.

Figure 5.12: Organizational knowledge creation (bottom-up model).

This principle stresses the collaboration of a group in knowledge creation, and the rational use of arational, subjective aspects in the organizational process of knowledge creation.

Organizational knowledge creation (top-down model) Gasson (2004) analyzed possible transitions between the same four nodes (*but group knowledge is called* shared knowledge *while individual knowledge is called* distributed knowledge) in the organizational culture of a Western company. The transitions have a different character and go in the direction opposite to the above model. See Figure 5.13.

Figure 5.13: Organizational knowledge creation (top-down model).

These models will be treated in detail in Chapter 6. The former one is in fact the *SECI spiral*, which was the key academic factor leading to the establishment of the School of Knowledge Science.

Academic knowledge creation (social science)   This is a model of individual knowledge creation supported by participation in a group. The process starts with the transition *Enlightenment*. This transition can give bigger or smaller insights: starting with small ideas and revelations up to basic, revolutionary enlightenments. This depends on the personality of the researcher, his/her preparation, and on using the special techniques supporting gestation. See Figure 5.14.

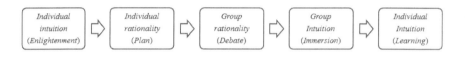

Figure 5.14: Academic knowledge creation (social science).

When this transition gives an individual researcher a novel idea—whether small or big—he/she usually rationalizes it and overcomes any difficulties related to finding words for new concepts, as in the case of the transition *Externalization*. The next transition is related to the presentation of this idea for a critical discussion with his/her colleagues. This is an extremely important transition from individual rationality to group rationality called *Debate*.

Academic knowledge creation (natural science)   It is necessary to extend the former spiral to represent the situation, where the verification of a new idea occurs not through *Debate* but through *Experiment*. The transition *Experiment* simply represents experimental research: verification of an idea, not necessarily of an already formed theory. See Figure 5.15.

Figure 5.15: Academic knowledge creation (natural science).

These models will also be explained in detail in Chapter 6.

5.3.3 An approach from systems science

Developing knowledge synthesis methodologies, methods, and tools is most important for knowledge science, using a variety of knowledge and media. When we face a complex problem, based on our experience-based knowledge, we make a plan to collect knowledge from the three dimensions and then synthesize the collected knowledge to obtain knowledge for problem-solving. See Figure 5.16.

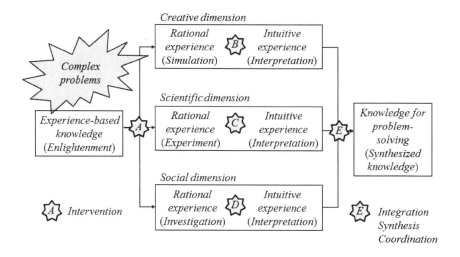

Figure 5.16: An approach from systems science.

In ordinary cases, we usually carry out one research process, B, C, or D. But as a knowledge science researcher, we should add part A: intervention and E: synthesis, separate from B, C, and D. This is in fact a systems approach itself that focuses on knowledge creation; however, the presence of the creative dimension is an innovative and challenging idea for systems science.

Question 5.8   How have you been carrying out knowledge synthesis? Cite examples from your research (or study) at universities or graduate schools.

Question 5.9   What is knowledge science in your understanding, at this point? Note that there is no guarantee that the systems science approach is appropriate.

Knowledge science can.be regarded as
a new systems science in the knowledge-
based society, which focuses on
knowledge integration and creation.

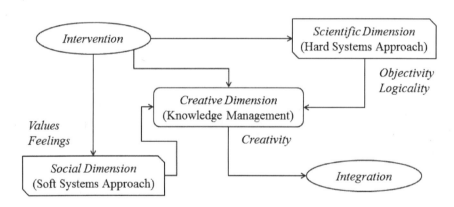

Knowledge science is a new level of systems science, which
integrates hard and soft systems approaches with creativity.

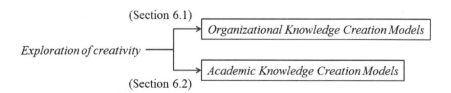

Chapter 6

Knowledge Creation Models

6.1 Organizational Knowledge Creation

The organizational knowledge creation theory developed by Nonaka and his colleagues provides a rational algorithmic-like recipe for generating new knowledge, using irrational, or even arational, abilities of the human mind and Oriental cultural features. But, see Figure 6.1.

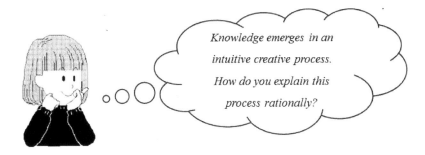

Knowledge emerges in an intuitive creative process. How do you explain this process rationally?

Figure 6.1: Can you explain your knowledge creation process rationally?

Where does creativity come from? How can we develop our creativity? Why can you generate a new idea? Do you have answers to these questions? If not, perhaps knowledge science has the answers.

Knowledge science is a discipline that deals with creativity.

6.1.1 Bottom-up Socialization, Externalization, Combination, Internalization (SECI) spiral

Nonaka and Takeuchi (1995) proposed a knowledge creation principle, which is revolutionary because it stresses steps leading to knowledge increase, based on group collaboration and on the rational use of irrational mind capabilities. See Figure 6.2.

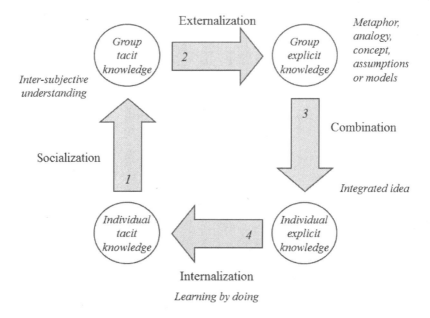

Figure 6.2: An organizational knowledge creation model.

1. *Socialization* is a process of sharing experiences and thereby creating tacit knowledge such as shared mental models and technical skills.

2. *Externalization* is a process of articulating tacit knowledge as explicit concepts that take the shape of metaphors, analogies, concepts, hypotheses, or models.

3. *Combination* is a process of linking explicit knowledge from which you can derive a knowledge system.

4. *Internalization* is a process of embodying explicit knowledge into tacit knowledge. It is closely related to learning by doing.

Socialization   *Socialization* is a process applicable to any organizational group. In Japanese society it can occur during an informal meeting of employees, drinking beer for relaxation and talking about anything, but including and often even concentrating on professional and current business problems of the organization.

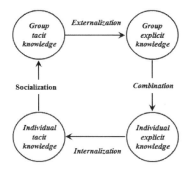

Both the emotive and intuitive parts of individual tacit knowledge are more easily exchanged this way than during the formal meetings typical in Western organizations.

From tacit to tacit   An individual can acquire tacit knowledge directly from others without using language. In the business setting, on-the-job training basically uses this principle. The following examples were given in Nonaka and Takeuchi (1995).

- "Honda's brainstorming camps: informal meetings for detailed discussion to solve difficult problems in a development project. The meetings were held outside the workplace, often at a resort inn where participants discuss difficult problems while drinking beer, sharing meals, and enjoying a hot spring together."

- "Matsushita Electric Industrial Company: in developing an automatic home bread-making machine, the engineers volunteered to apprentice themselves to a hotel's head baker, and they socialized the head baker's tacit knowledge through observation, imitation, and practice."

- "Socialization also occurs between product developers and customers. Interactions with customers before product development and after market introduction are a never-ending process of sharing tacit knowledge and creating ideas for improvement. The way NEC developed its first personal computer is a case in point."

Externalization   *Externalization* tries to express group tacit knowledge in words, to rationalize it. The examples given by Nonaka and Takeuchi (1995) imply the use of such new concepts or slogans, e.g., the concept of *twisting and stretching* summarizes the insight of tacit knowledge when working on a new home bread-making machine, or the slogan "man-maximum, machine-minimum" when working on the concept of a new car.

The Oriental tradition of trying to build consensus in group discussions might help in formulating new meanings.
Transition to both Socialization and Externalization might be difficult for Western culture, with its stress on individual achievements and on precise definitions.

From tacit to explicit   The externalization mode of knowledge conversion is typically seen in the process of concept creation and is triggered by dialogue or collective reflection. Examples shown in Table 6.1 are given in Nonaka and Takeuchi (1995).

Table 6.1: Examples of externalization.

Product (Company)	Metaphor/Analogy	Influence on Concept Creation
Honda City car (Honda)	Automobile evolution (metaphor) The sphere (analogy)	Suggestion of maximizing passenger space as the ultimate goal in auto development. Man-maximum, machine-minimum concept created. Suggestion of achieving maximum passenger space through minimizing surface. Tall and short car (Tall Boy) concept created.
Mini-Copier (Canon)	Aluminum beer can (analogy)	Suggestion of similarities between inexpensive aluminum beer can and photosensitive drum manufacture. Low-cost manufacturing process concept created
Breadmaker (Matsushita)	Hotel bread (metaphor) Osaka International Hotel head baker (analogy)	Suggestion of more delicious bread. Twist dough concept created.

Combination *Combination* occurs between the nodes of group explicit knowledge and individual explicit knowledge. However, the name of this stage suggests combining various elements of individual explicit knowledge with those of a group, into a rational synthesis together with some elements derived from the rational heritage of humanity.

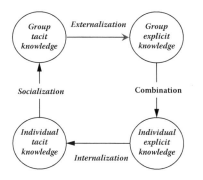

Combination is not a one-directional transition or a conversion from group explicit knowledge into individual explicit knowledge.

From explicit to explicit In the business context, the combination mode of knowledge conversion is most often seen when middle managers break down and operationalize a corporate vision, business concepts, or product concepts. At the top management level of an organization, the combination mode is realized when mid-range concepts (such as product concepts) are combined with, and integrated into, grand concepts (such as corporate vision) to generate a new meaning of the latter. Examples are given in Nonaka and Takeuchi (1995):

- "NEC's 'C&C' (computers and communications) concept sparked the development of the epoch-making PC-8000 personal computer, which was based on the mid-range concept of 'distributed processing'."

- "Canon's corporate policy, 'creation of an excellent company by transcending the camera business', led to the development of the Mini-Copier, which was developed with the mid-range product concept of 'easy maintenance'."

- "Mazda's grand vision, 'create new values and produce joyful driving', was realized in the new RX-7, 'an authentic sports car that provides an exciting and comfortable drive'."

Internalization   *Internalization* occurs between the nodes of individual explicit knowledge and individual tacit knowledge and means *learning by doing*. The rational synthetic instruction obtained by *Combination* must be tested in practice and new tacit knowledge should be created in this process.

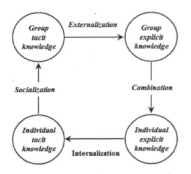

After completing four transition stages, knowledge is obviously increased rather than diminished and new perspectives obtained in each stage can only increase it; the increase might be small, but certainly will occur.

From explicit to tacit   When experiences through socialization, externalization, and combination are internalized into individuals' tacit knowledge bases in the form of shared mental models or technical know-how, they become valuable assets. Examples are also from Nonaka and Takeuchi (1995):

- "All members of the Honda City project team internalized their experiences of the late 1970s and made use of that know-how leading R&D projects in the company."

- "Expanding the scope of one's own experience is critical to internalization. For example, the Honda City project leader kept saying "Let's give it try" to encourage the team members' experimental spirit. The fact that the development team was cross-functional enabled its members to learn and internalize a breadth of development experiences beyond their own functional specialization."

But, is the SECI model accepted also in the West?

6.1.2 Top-down Objectives, Process, Expansion, Closure (OPEC) spiral

Gasson (2004) analyzed possible transitions between the same four nodes in the organizational culture of a Western company. Here group knowledge is called *shared knowledge* while individual knowledge is called *distributed knowledge*. Gasson argues that the transitions would have a different character and actually go in the opposite direction. See Figure 6.3.

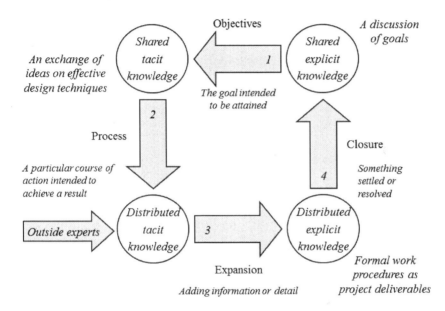

Figure 6.3: Another organizational knowledge creation model.

- The process starts with shared explicit knowledge, then a discussion of goals, and proceeds to shared tacit knowledge, leading to an exchange of ideas on effective design techniques.

- Once the group realizes that their shared tacit knowledge might not be sufficient, they specify expert advisors who should be invited to share their tacit knowledge with the group.

- Upon obtaining this additional expertise, the group comes back to individual, distributed explicit knowledge activity, by trying to define formal work procedures as project deliverables.

6.2   Academic Knowledge Creation

It is rather unlikely that the SECI spiral would result in a scientific break-through. Most probably, we would have to resort to other creative processes when addressing more fundamental research questions. The following models are given in Wierzbicki and Nakamori (2006).

6.2.1   Inter-subjective Enlightenment, Debate, Immersion, Selection (EDIS) spiral

This creative process is aimed not at organizational knowledge creation and management, but at individual creation of science and technology supported by participation in a group. See Figure 6.4.

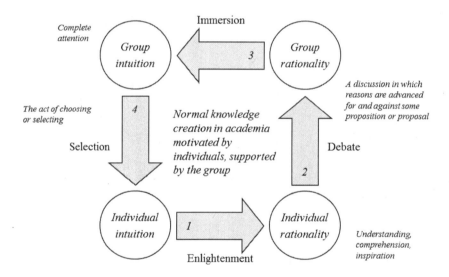

Figure 6.4: Inter-subjective EDIS spiral.

- The process starts with *Enlightenment*. This transition can give bigger or smaller insights, starting with small ideas and revelations up to foundational or revolutionary moments of enlightenment; this may depend on the personality of the researcher and his/her preparation. When this transition gives an individual researcher a novel idea, he/she usually rationalizes it and will overcome all the difficulties related to finding words for new concepts.

- The next transition is related to the presentation of an idea for critical discussion with colleagues. This is an extremely important transition from individual rationality to group rationality, called *Debate*.

- Scientific debate actually has two layers: one is verbal and rational, but after some time for reflection, we also derive intuitive conclusions from it. This is extremely important, and in fact the most difficult transition, called *Immersion* (of the results of debate in group intuition); it occurs from group rationality to group intuition.

- An individual researcher does not necessarily accept all the results of group intuition; he/she makes his/her own *Selection* in the transition from group intuition to individual intuition.

No knowledge is lost during all these transitions and each transition can add new perspectives, ideas, or insights, contributing to a deeper understanding in the next iteration of the process. Thus, this process guarantees knowledge creation, in smaller or bigger steps, depending on the situation. In some cases investigation is necessary before *Debate*, as shown in Figure 6.5, especially where a normal academic research process is involved.

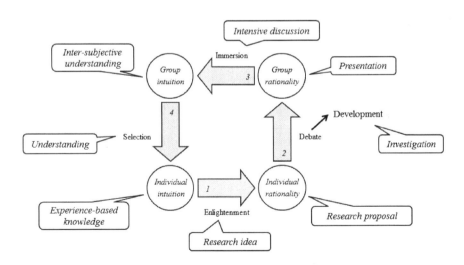

Figure 6.5: A small modification of the EDIS spiral.

6.2.2 Experimental Enlightenment, Experiment, Interpretation, Selection (EEIS) spiral

This model is related to normal knowledge creation in academia - a basic way to provide objectivity. See Figure 6.6.

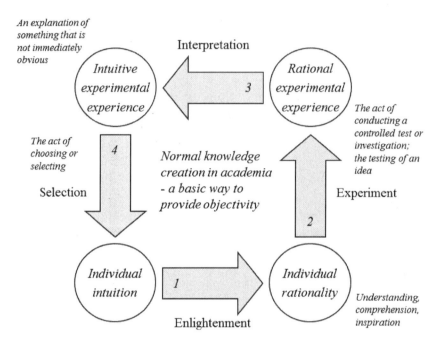

Figure 6.6: Experimental EEIS spiral.

- The transition *Experiment* simply represents experimental research — verification of an idea, not necessarily of an already formed theory.

- However, every researcher with experimental experience knows that raw experimental data does not provide much new knowledge; *Interpretation* is necessary in the sense of the immersion of experimental data in the intuition of the researcher, which is based on his/her experience.

- Here also an intuitive *Selection* follows, this time denoting choice of those aspects of data that have the biggest impact on the development of creative ideas.

6.2.3 Double EDIS-EEIS spiral

By combining the EDIS and EEIS spirals, we can achieve *a synthesis of inter-subjective and objective knowledge creation and verification.* Each loop of the double spiral might be repeated many times and the process can start at any node or transition. See Figure 6.7.

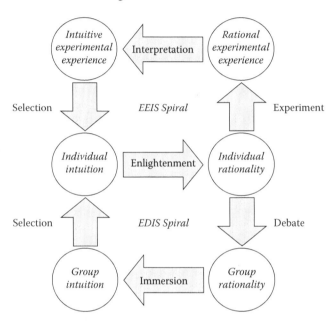

Figure 6.7: Double EDIS-EEIS spiral.

- An individual researcher reflects on perceived problems and gets an intuitive idea in the transition of *Enlightenment.*

- Then he/she has a choice of how to test this idea—through *Experiment* or through *Debate.*

- If *Experiment* is chosen, *Interpretation* of experimental results and *Selection* of conclusions follow.

- Thereafter, the researcher might want to present the idea to his/her colleagues and *Debate. Immersion* in group intuition and another *Selection* of suggestions follow.

6.2.4 Hermeneutic Enlightenment, Analysis, Immersion, Reflection (EAIR) spiral

This spiral describes the relationship of a researcher—a knowledgeable subject—to the object of his/her study represented by historical or literary texts, objects of art, etc. This spiral consists of four nodes: individual intuition, individual rationality, rational object perception, and intuitive object perception. See Figure 6.8.

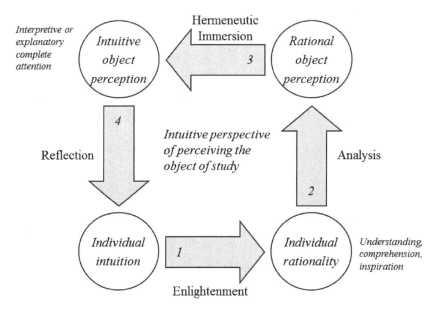

Figure 6.8: Hermeneutic EAIR spiral.

- *Analysis* brings rational perception to objects of study. However, this perception is not sufficient for a full understanding of objects.

- The researcher must therefore immerse this perception in an intuitive understanding of tradition, indicated by *Hermeneutic Immersion* that might be one of two types: *Critical* or *Integrated*.

- This immersion helps to achieve a deep *Reflection*, enriching individual intuition and leading to *Enlightenment*—new ideas about the object being studied.

6.2.5 Triple helix model

In the usual examples, and especially in interdisciplinary research, we use these three spirals in a mutually complementary manner. Three spirals can be represented together as a triple helix of normal knowledge creation:[1]

- *Hermeneutic EAIR Spiral*—searching through the rational heritage of humanity and reflecting on the object of study.

- *Inter-subjective EDIS Spiral*—debating ideas obtained from other spirals or through any other source of *Enlightenment*.

- *Experimental EEIS Spiral*—verification and objectification of ideas through experiments.

See Figure 6.9. The triple helix model is *isomorphic* with the *i*-System.

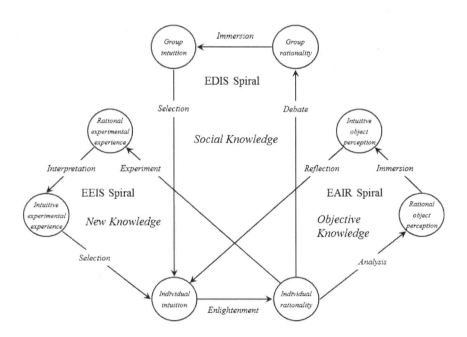

Figure 6.9: Triple helix model of normal knowledge creation.

---

[1]Tian et al. (2009) applied this triple helix model to the evaluation of knowledge management and knowledge creation in academia.

6.3 Application to Academic Research Evaluation

Here we combine the ideas of the *i*-System and the triple helix model, to evaluate knowledge creation processes and environments in research institutes or universities. See Figure 6.10, where Ai/Bi will be explained later.

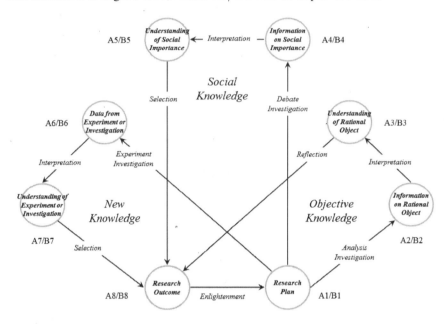

Figure 6.10: An evaluation model for knowledge creation in academia.

- Research is a continuous process by a scientist (even a graduate student must repeat the above three spirals to write a good thesis). In writing a research plan, we need enlightenment or inspiration based on previous research, including other researchers' outcomes.

- In each spiral, we distinguish knowledge from information, because research is often divided into activities to collect information and activities to interpret information.

- The justification principles of obtained knowledge are *novelty*, *usefulness*, and *logicality*, in the respective spirals for seeking objective, social, and new knowledge.

Checklist 1: On growth of students For each question below, the subject is asked to answer his/her satisfaction level and importance level for his/her research life:

Dissatisfied 1 2 3 4 5 Satisfied

Unimportant 1 2 3 4 5 Important

A1: Are you satisfied with your ability to plan your research and to explain your research activities cogently?

A2: Do you think that your investigation of the disciplinary status and perspective related to your research plan is sufficient and gives satisfactory results?

A3: When you have a research result or research idea, is it easy for you to explain its disciplinary importance (validity, uniqueness, etc.)? Do you have the confidence and intuition to explain this importance?

A4: Do you think that your investigation of the social importance of your research topic (social contribution, ripple effect, etc.) is sufficient and gives satisfactory results?

A5: When you have a research result or research idea, is it easy for you to explain its social importance? Do you have the confidence and intuition to explain this importance?

A6: Do you think that your abilities in carrying out experiments (or investigation, data analysis, etc.) are adequate and give satisfactory results?

A7: When you have experimental results or data, is it easy for you to interpret them, derive conclusions, explain their importance? Do you have the confidence and intuition to interpret them and explain their importance?

A8: When you have a new research result, is it easy to understand its overall (disciplinary, social, experimental) importance and its implications for further, new research? Do you have confidence and intuition to discover new research topics, to design new experiments, formulate new ideas?

Checklist 2: On research environments For each question below, the subject is asked to rate the sufficiency level and the necessity level of his/her research environments:

Insufficient 1 2 3 4 5 Sufficient

Unnecessary 1 2 3 4 5 Necessary

B1: When preparing a research plan, do you receive sufficient guidance from your supervisors or senior colleagues? If not, is it necessary?

B2: Is the availability of written research materials for your research topic (books, papers, research results of your supervisors and senior colleagues, both in print and electronic formats) sufficient in your lab and at the university? If not, is it necessary?

B3: Are discussions and guidance related to the disciplinary status of your research sufficient in your lab? If not, is it necessary?

B4: Is the information related to the social importance of your research sufficient in your lab and at the university? If not, is it necessary?

B5: Are discussions and guidance related to the social importance of your research sufficient in your lab? If not, is it necessary?

B6: Are the equipment and funds for carrying out experiments (investigation, data analysis, etc.) sufficient in your lab? If not, is it necessary?

B7: Are discussions and guidance related to experiments (investigation, data analysis, etc.) and their results sufficient in your lab? If not, is it necessary?

B8: Are discussions and guidance related to summing up your research and designing new research topics sufficient in your lab? If not, is it necessary?

Checklist 1 is related to questions about the abilities and actions of students, while Checklist 2 corresponds to the research environments or guidance systems.

Survey We carried out a survey among the graduate students in the School of Materials Science at Japan Advanced Institute of Science and Technology. The number of graduate students who answered the questionnaires was 103 including 33 doctoral students. The result is summarized as follows (Kikuchi et al., 2007).

Research ability First we asked graduate students about their capabilities and the importance to them of such abilities. Figure 6.11 shows the average answers of students, where the left-hand side shows the capabilities while the right-hand side shows the importance.

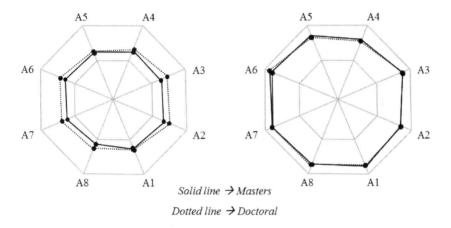

Solid line → *Masters*
Dotted line → *Doctoral*

Figure 6.11: Self-evaluation (left) and importance (right) of capabilities.

- The left-hand side shows that doctoral course students are superior to master's course students in almost all capabilities. On the other hand, the right-hand side shows that they consider these abilities very important for their research life.

- The correlation coefficients between the item A8 and A1, A2, and A6 are relatively high in the data of the doctoral course students. These items (planning research rationally, collecting information about research objects, and collecting experimental data rationally) influence a good research outcome.

Research environments   We then asked graduate students about the research environments including infrastructure and the guidance received from their supervisors. Figure 6.12 shows the average answers of students, where the left-hand side shows the evaluation of the environments while the right-hand side shows the necessity of such environments.

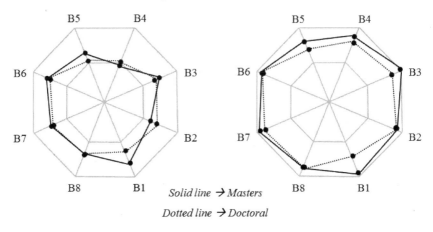

Figure 6.12: Evaluation (left) and necessity (right) of environments.

- From the left-hand side we see that the values of B1 and B6 are very high, which means that the support for research planning and in getting data from experiments is very good.

- But support for understanding the social importance of research is weak. From the right-hand side, we can see that students require more help in doing research in almost all aspects.

- From the correlation analysis, we found that B8 has relatively high correlations with B3, B5, and B7. This means that the acts of knowing at *Intelligence*, *Involvement*, and *Imagination* affect the act of knowing at *Integration*.

- Although the self-diagnosis with respect to investigation of the social significance of study is generally low, the students understood that paying full attention to this point is necessary for conducting good research.

6.4 Knowledge Creation Modeling: Exercise

Exercise 6.1

1. Give examples for which the SECI spiral best describes the knowledge creation processes.

2. Give examples for which the OPEC spiral best describes the knowledge creation processes.

3. Consider the process of research for a master's or doctoral thesis using the idea of the EDIS or EEIS model.

4. Consider the process of research for a master's or doctoral thesis using the idea of the Triple Helix model.

5. Try to draw a creative spiral, recalling the process of your own knowledge (ideas) creation. The nodes and transitions can be used as you wish.

6. Or, draw your research plan. How do you create knowledge in your research? Consider the names and meanings of the nodes and transitions.

Note We noticed that the *i*-System is almost identical with a combination of some knowledge creation models. Therefore, we can think of the *i*-System as being one of the general knowledge creation models. In the next chapter, we will try to justify this idea based on sociological interpretations and applications. Then, in the final chapter, we will try to develop a set of justification principles of knowledge created by the *i*-System.

The style of knowledge creation varies depending on the context. It is important to find your own model and refine it continuously.

Organizational Knowledge Creation (Cultural Difference)

| The West (Top-down) | ⟷ | The East (Bottom-up) |

Academic Knowledge Creation (Disciplinary Difference)

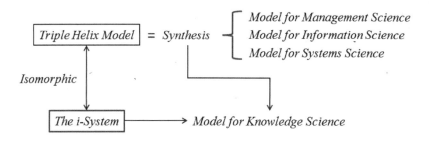

Chapter 7

Knowledge Synthesis or Construction

7.1 Sociological Interpretation

Knowledge is constructed and used by people in organizations and societies. This observation shows clearly that no generic model of knowledge is complete without sociological arguments. See Figure 7.1.

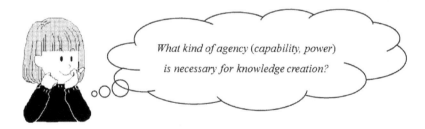

Figure 7.1: What kind of abilities do you have for knowledge creation?

The *i*-System should appear sufficiently sophisticated and inherently open to embracing sociological interpretations. In seeking a sociological underpinning, we will draw upon the structure-agency-action paradigm.[1]

---

[1]The main contributor for this elaboration is Zhichang Zhu, University of Hull, discussed in Nakamori and Zhu (2004).

Viewed through the *i*-System, knowledge is (re-)constructed by various actors, who are both constrained and enabled by structures, and realize and mobilize their agency. See Figure 7.2.

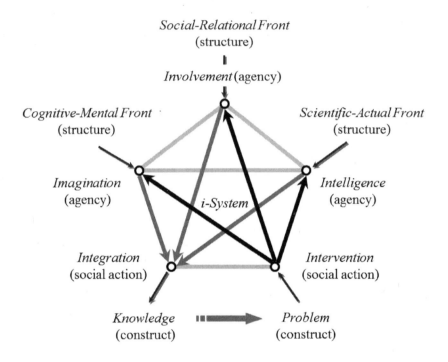

Figure 7.2: A sociological interpretation of the *i*-System.

- Actors are both constrained and enabled by structures that consist of a *scientific-actual*, a *social-relational* and a *cognitive-mental* front.

- They realize their agency and the agency of others, which can be differentiated as *Intelligence*, *Involvement*, and *Imagination* clusters.

- They might engage in *rational-inertial*, *arational-evaluative*, and *post-rational-projective actions* in pursuing sectional interests.

- Knowing (*integration*) and practice (*intervention*) are seen as being integrally connected, from which knowledge is emerging and embodied over time, *back* into structures and agency.

Figure 7.3 indicates the relationship between *Intervention* and *Integration*, which are actually one thing. If you want quality knowledge, the problem has to be well-defined. But to define a problem well, you must already have quality knowledge.

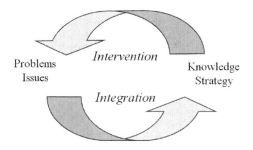

Figure 7.3: Knowing and action—a constructionist view.

Knowledge and action in Eastern thinking:

- Knowledge and action are but one, for purpose, and with consequences. (Wang Yang-ming, 1472–1529)

- Seeking knowledge by action, and for action. (Sun Yi-xian, 1866–1925)

And in the West:

- "Knowledge is 'knowing how', a capacity to perform/act in particular circumstances" (Ryle, 1949).

- "Our knowing is in our action" (Schon, 1983).

- "Knowledge is less about truth and reason and more about the practice of intervening knowledgeably and purposefully" (Spender, 1996).

- "All doing is knowing, and all knowing is doing" (Maturana and Varela, 1998).

- "The mutual constitution of knowing and practice. Knowing is an ongoing social accomplishment" (Orlikowski, 2002).

7.1.1   Structure complexity

Structure comprises the systemic, collective contexts and their underlying principles, which constrain and enable human action. With the *i*-System we see structure as consisting of *scientific-actual, cognitive-mental* and *social-relational* fronts. Here we choose the word *front*, not *dimension* or *domain*, because it conveys an indicative meaning in regard to our version of a constructivist view. See Figure 7.4, in which contents and the possible actions of actors in the respective fronts are shown.

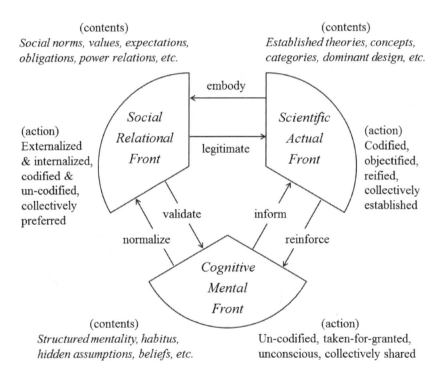

Figure 7.4: Three fronts as structures.

Structure comprises the systemic, collective contexts and their underlying principles, which both constrain and enable human action.

Scientific-actual front In the context of innovation, the scientific-actual front draws our attention to established scientific theories, concepts, and categories in natural science and engineering disciplines (Zhu, 2000).

- From a constructivist perspective, these are creations of human actions, never perfect or complete.

- Over time, however, either conceptual (e.g., engineering principles), virtual (e.g., customers' buying power), or material (laboratories and equipment) components in the scientific-actual front become "actual."

- That is, they are codified, normalized, objectified, and realized in and around organizations (Archer, 1995; Scott, 1995; Sewell, 1992).

Consequences of human action in relation to the scientific-actual front are largely predictable. Given such objectivity in this front, to act otherwise is suicidal.

> *The scientific-actual front provides testable answers to the empirical question, "What is it?"*

Social-relational front With the social-relational front we denote the mixture of social norms and values, expectations and obligations, codes of behavior and patterns of interaction, power relations and legitimacy — what is, or is not, appropriate, proper, acceptable, and right. In the social-relational front, we also find public policy and regulatory regimes in the context of technology innovation, for example, the way(s) research is organized, funded, evaluated, publicized, rewarded, and used. Some of the components in the social-relational front are codified, others are not; and there are others that never will be. The paradox here is as follows:

- The codified components do not always result in the expected actions in the absence of the backing from the uncodified components, since they are open to interpretation.

- Whereas the uncodified components, once neutralized and internalized, become powerfully informative and effectively regulative, since the actors know how they are expected to act and what they can expect from the others, based on their conduct.

The social-relational front is about the way people can be "rational" about desirable goals, appropriate means, and proper conduct (Parsons, 1950). In the inter-subjective social-relational front, to act otherwise is unviable due to expected social sanctions.

> *The social-relational front provides inductive answers to the normative question, "What should it be?"*

Cognitive-mental front   In the cognitive-mental front, actors live with deep-seated structured mentalities, collective rationality, frames of reference and beliefs, heuristic devices, habitus, styles, shared narratives and practices, common language and codes, collective, organizational and procedural memories, stored cognitive and action recipes, organizational routines, dominant logic, hidden assumptions, and paradigms.

- Like in the scientific-actual front, the components in the cognitive-mental front are rules and scripts for action.

- Unlike in the scientific-actual front, the components in the cognitive-mental front remain largely tacit and taken for granted (Sewell, 1992).

- If "seeing is believing" is the ideal in the scientific-actual front, "believing is seeing" is what happens in the cognitive-mental front.

From a constructivist perspective, the components of the cognitive-mental front, like those in other fronts, are man-made creations. Once in place, however, they become deeply ingrained, relatively inertial but largely invisible. "This is the way things should be seen and done" (Arthur, 1988; David, 1985; North, 1990)—nobody talking or even thinking about it but everybody sharing it and acting accordingly. Although a subjective matter, 'to act otherwise' is difficult because it is an inconceivable option.

> *The cognitive-mental front provides unproblematic answers to the hidden question, "What should it be?"*

Constructionist view We suggest that the actors are embedded in the structure consisting of the three fronts, in the sense that they take action within the conditioning, and with the support, of the fronts. The efficacy, efficiency and quality of a technology innovation will always be bound by Mother Nature and other actual factors, and thus it is essential for researchers to master the scientific-actual components. Yet innovation is in its nature uncertain, ambiguous, and risky.

- No ready, complete, or clear answer is available in advance in the *scientific-actual front*.

- The *cognitive-mental front* is therefore indispensable for shaping innovations by providing action strategies.

- Furthermore, the resources available, the forms, functionalities, and evaluation routines adopted, and hence the eventual success/failure of the innovation are all dealt with by the wider communities, stakeholders, and society, which is the subject matter of the *social-relational front*.

Our constructivist view holds that it is not a choice to be or not to be embedded: actors and innovations are born into the fronts, which set the limits of what is workable, conceivable, and acceptable, and therefore set tendencies and parameters along which other choices are made. We, with the *i*-System, posit that:

- It is the complexity inherent within and between the fronts that provides emergent opportunities for transformative innovation (as well as for recursive preservation).

- By complexity we mean the oppositional relations with coherence, complementarities, endurance, and profoundness at the one end, and plurality, tensions, uncertainty, and ambiguity at the other.

> *Innovation is not solely the functionality of structure. Structure complexity only provides emerging possibilities. It is human agency that makes the difference, to which we now turn.*

7.1.2 Agency complexity

The *i*-System differentiates human agency into *Intelligence*, *Imagination*, and *Involvement* clusters, so that agency can be understood in an organized way, and not treated like a black box. See Figure 7.5, in which contents and their natures in respective agencies are shown.[2]

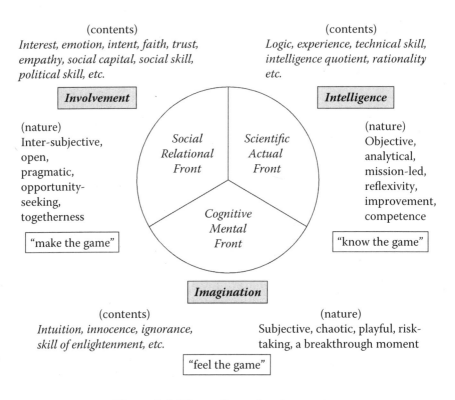

Figure 7.5: Three-dimensional agencies.

> *Agency is the capability with which actors, who are socio-technologically embedded, reproduce and transform the world.*

---

[2]Note that this book defines *agency* as the capability to act, instead of defining it as the act itself, in order to connect the *i*-System to social theories.

Intelligence By *intelligence* we mean the intellectual faculty and capability of the actors: experience, technical skill, functional expertise, intelligence quotient, etc. The vocabulary related to intelligence addresses logic, rationality, objectivity, observation, monitoring, and reflexivity. The accumulation and application of intelligence are mission led and rationale focused, discipline and paradigm bound, confined within the boundaries of "normal science" (Kuhn, 1970), which leads to "knowing the game" and incremental improvement of the components.

> *Seeing intelligence as inertial and paradigm bound, the i-System does not regard intelligence as negative per se. Rather, to the i-System, intelligence is indispensable for creativity.*

As Polanyi (1958) puts it, "science is operated by the skill of the scientist and it is through the exercise of this skill that he shapes his scientific knowledge." Following Sewell (1992), we see the search for intelligence as a process of *transposition*, and actors apply and extend codified rules and procedures "to a wide and not fully predictable range of cases outside the context in which they are initially learned." Intelligence becomes a liability to innovation only when it blocks actors from seeing alternatives.

Involvement *Involvement* is the cluster in human agency that consists of interest, faith, emotion, and passion, which is intrinsically related to intentionality and "habits of the heart," as well as the social capital, social skill, and political skill that make intentionality and *the heart* felt. We talk about the quality of involvement in terms of empathy, trust, care, togetherness, and openness, as well as competition, fragmentation, conflict, politics, manipulation, domination, and hegemony.

> *Involvement is inter-subjective, communicative, discursive, and pragmatic in nature, and therefore in the realm of arationality.*

As a human agency, involvement can produce managerial and institutional effects, particularly in dealing with the social-relational front, in that it can help or hamper researchers' efforts to "make the game."

Imagination   In the *imagination* cluster we uncover intuition, innocence, ignorance, skill of enlightenment, and post-rationality, which inspires a vocabulary including feeling the game, playful, fun, chaotic, illogic, forgetting, up-setting, competency-destroying, knowledge-obsolescence, and risk-taking. For this we turn to naturalist Taoism, transcendental Zen Buddhism and pragmatic Confucianism. Zhuang Zi, the Taoist sage, famously proclaims that "great knowledge is like a child's ignorance." He distinguishes three kinds of knowledge:

- pre-rationality (child's knowledge, or *primary ignorance*),

- rationality (adult's knowledge, i.e., *great artifice*, which denotes established theories, concepts, categories, "normal science" and associated findings), and

- post-rationality (absence of knowledge, or *true knowledge*, i.e., the knowledge of the "True-man").

To attain true knowledge, the methodology of Taoism is *forgetting* and that of Zen Buddhism is *emptying the mind*, *sudden enlightenment*, and *liberation*. Even the rationalist Confucius has reported that "the master had no foregone conclusions, no arbitrary pre-determinations, no obstinacy, and no egoism." Confucianism, like Taoism and Zen Buddhism, emphasizes synthesizing and transcending rationality.

> *Intuition, defined here, is the natural synthesis of the outcomes of moral maturation, intensive observation, direct experience, and persistent intellectual effort.*

Emergence-theoretical perspective   To the *i*-System, while structure complexity provides emergent opportunities for innovation, agency complexity provides emergent possibilities for actors to exploit those opportunities. To understand how actors deploy their agency to exploit innovation opportunities, we need further to investigate in a systemic way what actors do when engaging with potential opportunities.

7.1.3 Action complexity

The *i*-System adopts a *chordal triad*, a construct that disaggregates social action to *rational-inertial*, *arational-evaluative*, and *post-rational-projective* dimensions along which actors contextualize past experiences, evaluate them socially, and imagine alternatives for the future. See Figure 7.6.

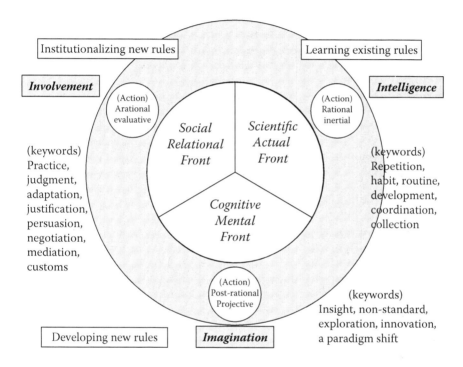

Figure 7.6: Actors' actions in three fronts.

This idea has deep roots in Western philosophy and notably American pragmatism[3] and Continental phenomenology.[4]

---

[3]*Pragmatism* is a philosophical tradition centered on the linking of practice and theory. It describes a process where theory is extracted from practice, and applied back to practice to form what is called *intelligent practice*.

[4]*Phenomenology*, in Husserl's conception, is primarily concerned with the systematic reflection on, and study of the structures of, consciousness and the phenomena that appear in acts of consciousness.

If we regard the *i*-System as a systems methodology to solve complex social problems, we usually proceed as follows:

1. The rational-inertial dimension refers to the selective reactivation, in the face of emergent situations, of past patterns of thought and action as ingrained in the actors' intelligence.

2. The arational-evaluative dimension concerns feasible actions that adjust past dispositions and bring projected trajectories down to earth.

3. The post-rational-projective dimension involves imaginative generation of possible trajectories for the future.

However, since we are offering the *i*-System as a knowledge synthesizing model, we treat these three dimensions as an unordered parallel relationship.

Rational-inertial dimension   As new contingencies never completely coincide with learned types, patterns, and rules, actors apply their knowledge to new circumstances. To implement this, actors intentionally appropriate their existing knowledge so as to get things done. However, it is too apparent that inertial action alone is not sufficient for effective knowledge construction, although it allows for the use of supporting resources for that purpose.

Arational-evaluative dimension   The practical-evaluative dimension concerns feasible actions that adjust past dispositions and bring projected trajectories down to earth. It is a real-time situated exercise of judgments and decisions, given the usually competing ends–means relations, resources constraints, and not-known-in-advance action consequences.

Post-rational-projective dimension   When situations unfold such that the problems cannot satisfactorily be resolved by habitual thought and action, actors construct changing images and options of where they think they are going, where they want to go, and how they can get there, in various culturally embedded ways. In projecting the future, actors deploy mainly their imagination agency, and engage in several important processes. Crucial and beneficial as they are, effective projection is difficult to come by because it disturbs the status quo of all of the fronts — scientific-actual, social-relational, and cognitive-mental. Being creative, projection is at the same time destructive (Schumpeter, 1934).

7.1.4 Universality of the *i*-System

The multifront, multicluster, multidimension of the *i*-System's conception can be seen as rooted in Confucianism, a life philosophy of the Chinese and the Japanese. Although mainly concerned with epistemology (the study of knowledge) and methodology (the study of methods) in social life, and lacking sufficient interest in metaphysics, the Confucian teaching does, in our view, imply a latent view of reality. That is, reality as a complex web of relations: relation with nature, relation with the mind, relation with humans (others). The *i*-System conceptions of structure, agency, and action are to us, therefore all informed by these chordal triad relations. We find interesting affinities between the *i*-System's conceptual multiplicity with the following, though the terminology is diverse and slightly confusing:

- Giddens's (1979) facility, interpretive scheme, norm modalities and signification, domination, legitimation structures;

- Habermas's (1972) three worlds and corresponding human interests and knowledge;

- Archer's (1995) structural and cultural conditions;

- Scott's (1995) regulative, cognitive, and normative pillars of institutions;

- Child's (1997) material, cognitive, and relational structures;

- Nahapiet and Ghoshal's (1998) structural, cognitive, and relational dimensions of social/intellectual capital; and

- Garud and Rappa's (1994) three basic definitions of technology: technology as objective artifacts, as subjective beliefs, and as legitimized normative evaluation routines.

This convinces us that the *i*-System's sociological background is on the one hand localized and culturally bound, which is manifested in its emphasis on, for example, ignorance, emotion, and dialectics in terms of complementary oppositions rather than of the thesis-antithesis-synthesis grand order. But it is on the other hand universal, on the grounds that the *i*-System shares many similar concerns, values and conceptual patterns, such as chordal triad conceptions, with its Western counterparts.

7.2   Application of Interdisciplinary Integration

Demand forecasting is an important management skill at grocery supermarkets that deal with perishable foods. Because perishable foods generally have a short shelf life and are difficult to keep fresh, errors in demand forecasting result in excess product orders or production, which lead to added preservation/freshness management and personnel costs, and losses due to waste and forced discounting. On the other hand, demand in excess of stock results in losing opportunity and customer trust.

Problem definition   The problem definition and the knowledge that we seek are shown in Figure 7.7, and the tasks in the respective fronts are described below.

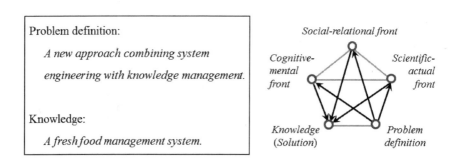

Figure 7.7: The purpose of developing a fresh food management system.

- *Scientific-actual front*: Construct a demand forecasting model that forecasts the sales amount under certain conditions based on past sales data. (In fact, a system was developed based on the outcome of the social-relational front and the cognitive-mental front.)

- *Social-relational front*: Collect and analyze consumers' opinions when purchasing products, which is not reflected in the data.

- *Cognitive-mental front*: Survey and analyze issues identified by the manager, who has the authority to decide the amount and types of products to be produced, and the management method related to daily sales.

Why an integrated approach? Studies on demand forecasting have been developed in various fields, such as systems engineering and operations research, utilizing the data from information systems such as POS (point of sale) and FSP (frequent shoppers program). However, the theoretical research does not seem to yield very impressive results at the actual retail business work site. The points listed below may explain why research achievements in demand forecasting may not function well in real-world situations:

1. Demand forecasting with mathematical and statistical models is not as flexible as human judgment or analysis.

2. Since it is difficult to obtain all the data necessary to calculate demand, the forecast is inevitably inaccurate.

3. Products are switched with high frequency, so the amount of data for relatively new products is small, lowering forecast accuracy.

We therefore need to develop a demand forecasting system by integrating knowledge from the scientific-actual, social-relational, and cognitive-mental fronts. See Figure 7.8.

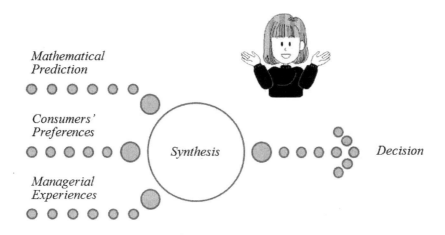

Figure 7.8: The necessity of knowledge integration.

7.2.1   Scientific-actual front

To develop a demand forecasting model, we took the following approach:

- Approach:

    - Group days that have similar conditions.

    - Predict the impact of conditions on the sales amount.

- Method:

    - Treat each conditional field as a categorical attribute:

        * e.g., weather = sunny, rainy, cloudy, snowy

    - Perform clustering on categorical attributes.

    - Show the potential sales amount under the respective conditions.

Data was grouped into clusters with the same conditions, and the distributions of sales amounts were extracted under those conditions. The clustering algorithm is introduced in the appendix of this chapter. See Figure 7.9.

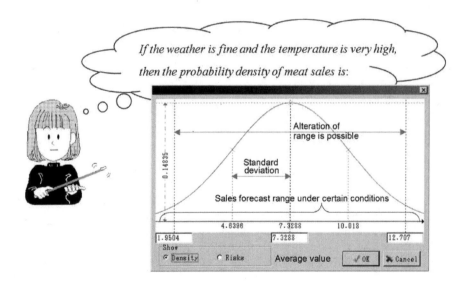

Figure 7.9: A demand forecasting model.

Risk analysis To determine the amount of each food to prepare, we use the risk analysis:

- Approach:

  - Draw the waste and opportunity loss distribution curves.

  - Evaluate the risk of respective production targets.

- Method:

  - Introduce two measures related to the risk:
    * Waste loss (loss due to discarding unsold product)
    * Opportunity loss (loss due to being sold out)
  - Find the optimal point to balance the two losses.

The expectation curves of waste loss and opportunity loss are drawn depending on conditions such as weather, temperature, events, etc. The idea of risk analysis is introduced in the appendix of this chapter. See Figure 7.10.

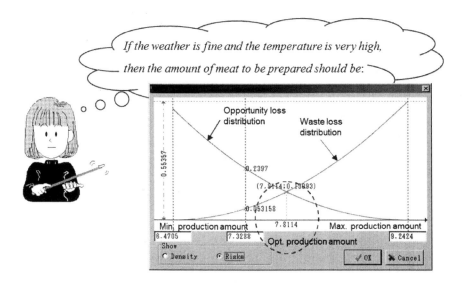

Figure 7.10: Decision making using risk analysis.

7.2.2   Social-relational front

In order to investigate the purchasing behaviors of consumers, we conducted a web survey of 1,000 people from 16 to 69 years of age living in Tokyo, who make prepared food purchases two or more times a week at supermarkets.

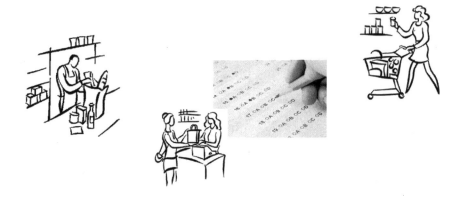

From the survey we gained important knowledge, for instance:

- Almost all age groups purchase prepared food mainly on the weekend.

- All age groups purchase prepared food during the dinner time period, from 16:00 to 19:00. However, the age group of those in their 60s showed a tendency to purchase a lot of prepared food in the lunch time period, from 11:00 to 13:00.

- In terms of the prepared food products frequently purchased, there were similar trends for all age groups. The type of prepared food product with a high purchase frequency was fried foods.

- Factors such as seasoning, the balance between price and amount, past purchasing experience, and appearance of cleanliness were of high importance.

- Factors such as weather, and details such as holidays, television shows, and regional events, were of low importance. This is contrary to the ideas of prepared food department managers.

7.2.3   Cognitive-mental front

We conducted an interview survey with ten prepared food department managers and gathered managerial knowledge for the prepared food department. We used the semistructured interview method. While a structured interview has a formalized, limited set of questions, a semistructured interview is flexible, allowing new questions to be brought up during the interview as a result of what the interviewee says (see, for instance, Lindlof and Taylor, 2002).

- The interview began with simple questions consisting of the basic information of the survey participant (years of work experience and position) and basic store information (such as the number of employees, the average number of customers, and the business hours).

- It gradually progressed to the core survey items (questions covering decision-making methods related to the amount and types of prepared food products, preparation in regard to prepared food production, etc.).

- Finally, we asked about decision making concerning prepared food products (measures taken against opportunity loss and waste loss, and issues connected to demand forecasting).

After making a transcript of the information from the interview, we structured the knowledge using the *affinity diagram*, which allows for large numbers of ideas to be sorted into groups, based on their natural relationships, for review and analysis (see, for instance, Britz et al., 2000). See Figure 7.11.

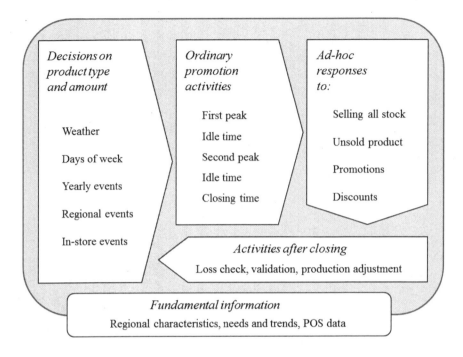

Figure 7.11: Outline of management methods.

We will not go into detail on the recommended activities, however, the ad hoc responses can be summarized as follows:

- *Selling all stock*: If most prepared foods seem to be sold out, produce more prepared foods quickly; if only one specific type of prepared food is sold out, produce more of that food.

- *Unsold product*: It occurs when large amounts of prepared foods are unsold due to inability to store leftover raw materials, weather reports are wrong, or stores that are in competition start limited-period sales.

- *Promotions*: For instance, tasting/samples, in-store announcements, sample menu display, changes in products or product combinations.

- *Discounts*: Start discounting, looking at movement of stock, and raise the discount rate gradually over time.

7.2.4 Interdisciplinary knowledge synthesis

A perfect demand prediction is impossible due to several reasons:

- Since an order must be placed to purchase any ingredients, the actual demand forecast must be carried out one week to ten days before.

- There is a need to modify the forecast, such as if there will be an event in the neighborhood.

- The company headquarters often plans volume sales of a particular food, which causes a modification in the forecast of similar foods.

- Modification is also needed for special situations such as a typhoon.

We therefore need to synthesize knowledge from the three fronts as shown in Figure 7.12.

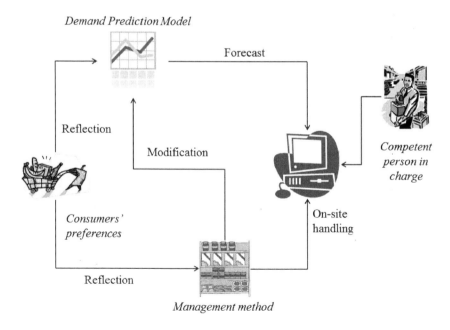

Figure 7.12: Knowledge synthesis and decision making.

Ideas of synthesis

- When forecasting the demand:

  - Change the minimum and maximum amount of production, taking into account managerial knowledge.

  - Shift the optimum amount of production, taking into account managerial knowledge and consumer purchasing behavior.

- On-site:

  - While looking at the sales, refrain from making more product, or else promote sales.

> *In any case, an experienced, competent person in charge is necessary to make decisions: a knowledge coordinator.*

7.3   The *i*-System: Exercise

Exercise 7.1

1. Consider applications of the *i*-System for collecting and synthesizing knowledge to solve any complex problems.

2. Or, consider applications of the *i*-System in compiling a research plan for a master's or doctoral thesis:

   - What should be done in each dimension or front?

   - What kind of ability is necessary in each dimension or front?

   - What type of action is required in each dimension or front?

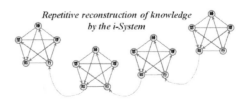

Repetitive reconstruction of knowledge by the i-System

7.4 Appendix on Risk Analysis

Let F be the distribution of sales amount of certain foods under certain conditions such as sunny weather, hot weather, weekends, etc., and let f be the probability density function of F. See Figure 7.13.

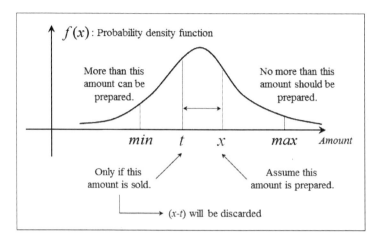

Figure 7.13: An example of probability density function of sales amount.

Let x denote the production target of this prepared food, that is, x is the point at which to evaluate the risk.

- The expected value of the waste loss is given by

$$L_w(x) = \int_{\min}^{x} (x - t)f(t)dt.$$

- The expected value of the opportunity loss is given by

$$L_o(x) = \int_{x}^{\max} (t - x)f(t)dt.$$

- The balance point x_0 is given by

$$x_0 = \frac{\displaystyle\int_{\min}^{\max} tf(t)dt}{\displaystyle\int_{\min}^{\max} f(t)dt}.$$

- If $\min = -\infty$, $\max = +\infty$, then $x_0 =$ the average of f.

7.5   Appendix on Categorical Clustering Methods

The following is a summary of clustering of the k-means paradigm:

- k-means method for numerical data (MacQueen, 1967)

  1. Determine arbitrarily k initial centers of groups.
  2. Assign each object to the nearest group.
  3. Recalculate the center of clusters.
  4. Repeat the above two steps until no change occurs.

- k-mode method for categorical data (Huang, 1998)

  1. Introduce a non-similarity measure between objects.
  2. Use the mode instead of the average.
  3. Update the mode by the method based on the frequency.

- k-representative method for categorical data (Ohn Mar San et al., 2003)

  1. Introduce a new definition of the cluster center.
  2. Define the center instead of the mode, taking vagueness into consideration.
  3. Redefine the distances between objects and clusters.

k-representative method

- Attention to a cluster C consisting of categorical objects.
- Suppose that each object X has m categorical attributes (e.g., weather, season, etc.):

$$X = (x_1, x_2, \cdots, x_m), \quad x_i \in D_i.$$

- Determine a representative of the cluster C as

$$Q(C) = (P_1, P_2, \cdots, P_m),$$

where P_i is the discrete probability distribution on D_i.

- Define the dissimilarity between X and C by

$$d(X, Q(C)) = \sum_{i=1}^{m} (1 - P_i(x_i)).$$

Example 7.1

- Objects in a cluster C are shown in Table 7.1:

    - Suppose that there are ten objects (business days).
    - Suppose that each object has two attributes.

    Table 7.1: Objects and attribute values in C.

Object	Weather	Temperature
1	sunny	hot
2	sunny	warm
3	cloudy	cold
4	rainy	cold
5	rainy	warm
6	cloudy	cold
7	rainy	warm
8	snowy	cold
9	cloudy	warm
10	cloudy	hot

- Representative $Q(C) = (P_1, P_2)$ for the cluster C (see Figure 7.14):

    - P_1: sunny 20%, cloudy 40%, rainy 30%, snowy 10%.
    - P_2: hot 20%, cold 40%, warm 40%

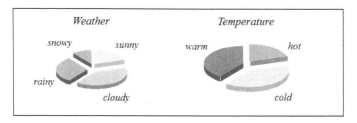

Figure 7.14: Example of cluster representatives.

- Dissimilarity between X=(sunny, warm) and the cluster C:

$$d(X, Q(C)) = (1 - 0.2) + (1 - 0.4) = 1.4.$$

Interdisciplinary integration cannot be
performed successfully without talented
people called knowledge coordinators.

From the sociological arguments:
Knowledge is constructed by actors who are both constrained and enabled by structures.

Agency → Capability that the actor should demonstrate in each social structure

- ✓ *Intelligence → Capability to collect and understand scientific knowledge*
- ✓ *Involvement → Capability to collect and prove social knowledge*
- ✓ *Imagination → Capability to generate and justify original knowledge*

From the constructionist view:
Knowledge is constructed rather than synthesized.

Agency → Capability that the actor should demonstrate in systemic intervention

- ✓ *Intervention → Capability to intervene in each social structure*
- ✓ *Integration → Capability to (re)construct systemic knowledge*

By comparison with social theories:
The locality as well as universality of the i-System has become clear.

Chapter 8

Knowledge Justification

8.1 Knowledge Discovery and Justification

Historically there have been many diverse attempts to understand how knowledge is created. Generally, until the last decade of the twentieth century we could distinguish two main schools of thinking. The groupings reflect an attitude that leads to the possibility of distinguishing the *context of knowledge discovery* and the *context of knowledge justification.*[1] See Figure 8.1.

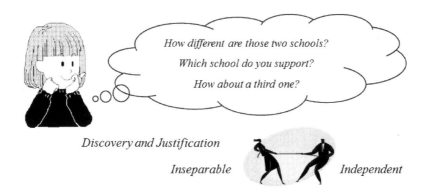

Figure 8.1: Which school do you support?

[1]The main contributor of this summary is Andrzej P. Wierzbicki, National Institute of Telecommunications, Poland. See Wierzbicki and Nakamori (2006, 2007a, 2007b).

The first school   The first school maintains that knowledge creation is essentially different from knowledge justification and validation, thus distinguishing the context of discovery and the context of justification. This school also maintains that creative abilities are irrational, intuitive, instinctive, and subconscious. This view is represented by many great thinkers of very diverse philosophical persuasions and disciplinary specialties.

- Bergson (1903) stressed the creative role of intuition, but understood it as an irrational, mystic force.

- Poincaré (1913) stressed the role of illumination or enlightenment in a creative process.

- Popper (1934) underlined logically the earlier conclusion of David Hume (1711–1776) that physical induction gives no guarantee of truth and thus postulated that new theories, obtained by irrational creativity, should be subject to falsification tests.

- Kuhn (1970) denied the possibility and rationality of falsification, but admitted that the new concepts that form the basis of a scientific revolution result from creative, irrational acts.

- Selye (1964) stressed the role of vision and intuition.

- Polanyi (1966) described creativity as the result of personal tacit knowledge, which contains instincts, myths, and intuition.

In addition to these:

Nietzsche (1844–1900) believed in the dominating role of irrational human will.

Brouwer (1881–1966) and Gödel (1906–1978), each for a different reason, believed that all mathematics is based on intuition.

Einstein (1879–1955), Heisenberg (1901–1976), Bohr (1885–1962), all stressed the diverse irrational aspects of creative acts.

Freud (1856–1939) explained creativity by subconscious instincts. Jung (1875–1961) explained creativity by myths in the collective unconscious.

The second school The second school keeps to the old interpretations of science as a result of empirical experience, induction, and logic, and refuses to see creative acts as irrational. This view is represented by many hard scientists. The opinion that science is the result of inductive reasoning and that creative acts can be perfectly logically explained has long been put forward in many publications, particularly by representatives of the hard sciences. Such a theory can be rationalized by maintaining that there is no distinction between the context of discovery and the context of justification, that there is only a joint creative process that can be perfectly logically planned, that intuition is only accumulated experience, and that revelation is only a revision of hidden assumptions.

The third approach In knowledge science, we are seeking a third way, in order to promote knowledge science as a discipline. See Figure 8.2.

Logical: *Knowledge is created inductively.*
(Logical empiricism)
(Hard sciences)

Context of discovery and justification

Illogical, or tacit: *Knowledge emerges from creative activities, which cannot be explained logically.* (Nietzsche, Bergson, Poincaré, Brouwer, Einstein, Heisenberg, Bohr, Freud, Jung, Popper, Kuhn, Polanyi)

Synthesis

The third approach=integrated: *Knowledge emerges from creative activities or intuitive (emotional) creative processes. However, the process can be analyzed rationally. The first example is the SECI model.*

Figure 8.2: The third approach to knowledge discovery and justification.

8.2   Philosophical Positions

The standards of testing theories belong to the episteme—the prevalent way
of creating and justifying knowledge which is characteristic of a given historical era or a cultural sphere (Foucault, 1972). Originally, *episteme* was
knowledge or science, but Foucault proposed to use this term as the knowledge framework of a certain period.

- At first he considered that a person's idea would follow his/her thinking
  system, but later he modified his approach and said that the episteme
  varies according to people's new knowledge.

As already discussed in Chapter 5, the episteme of industrial civilization,
called sometimes the *modern episteme*, was subjected to a destruction process, particularly visible in the last fifty years. This has led to a divergent
development of separate epistemes for the three cultural spheres: hard and
natural sciences, technology, and social sciences and humanities. See Figure
8.3.

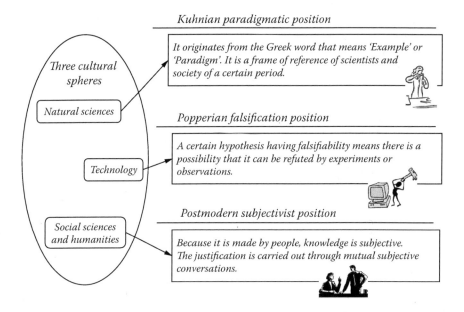

Figure 8.3: The epistemes in three cultural spheres.

8.2.1 Paradigmatism

The first block on the upper right in Figure 8.3 is related to the works of Kuhn (1962), and particularly his concept of scientific revolutions and a normal, paradigmatic development of science between these revolutions. Testing new normal scientific developments strongly depends on the paradigm:

> *Theories should fit observations or outcomes of empirical tests. Although theories that are consistent with the paradigm are welcome, theories that contradict the paradigm are rejected, even if they would better fit observations or empirical outcomes.*

Testing hard science theories

- Testing and justifying scientific theories (mostly hard science theories) has been the main theme of epistemology in the twentieth century. Hard and natural sciences follow established paradigms in their normal development and change paradigms rarely during scientific revolutions.

- Recently, Król (2007) continued the analysis of a core of axioms and has shown that the interpretation of the truth of such axioms constitutes an intuitive hermeneutic horizon.

- A theory in basic, hard and natural science is first tested concerning its logical consistency with the paradigm of the scientific field; but, even more importantly, we evaluate the truth of this consistency according to the hermeneutic horizon of this field.

8.2.2 Falsificationism

Opposing the Kuhnian paradigmatic position was the Popperian falsification position or falsificationism. In Popper (1934), the main goal was to show the logical inconsistency of the positivistic concept of induction from facts. It was suggested that:

> *Theories might be formed by inadequate induction, or by any other intuitive illumination. Therefore, they should be falsified, and subjected to tests aimed at disproving them.*

Testing knowledge creation in technology

- Intuitive, artistic creation of tools implies that we cannot fully formalize this activity: No matter what quality control we apply in technological processes, a new tool might always be dangerous. Therefore, tools are tested rigorously, and subjected to destructive tests.

- Thus technology, as opposed to science, follows falsificationism in its everyday practice.

- Unlike rational hard science, technology is more intuitive in its creation of knowledge, and follows falsificationism rather than paradigmatism in knowledge justification.

8.2.3   Postmodern subjectivism

The most radical position with respect to knowledge justification developed in the social sciences and humanities together with postmodernism; we shall call it the *postmodern subjectivist position* or *postmodern subjectivism*. The dominant position in the social sciences, particularly in postmodern sociology of science, is that:

> *Knowledge is constructed by people, is thus subjective, and its justification occurs only through inter-subjective discourse.*

Testing social science theories

- Since individual, personal versions of truth depend on individual experience and local context, each participant of a social discourse has the right to insist on his/her own perspective. This has led to diverse paradigms in social sciences and humanities.

- In sociology, Jackson (2000) counts four groups of paradigms: functionalist, interpretive, emancipatory, and postmodern.

- A pluralistic worldview is important, but to preserve the best knowledge for future generations, we need principles of objectivity and self-critical attitudes even in a pluralistic world.

8.3 Discussion and Exercise

Even if interdisciplinary and philosophical theories have many aspects that are related either to social sciences or to hard and natural sciences, we cannot test them today from the perspective of only one scientific episteme. The testing of interdisciplinary and philosophical theories should thus include:

- A description and critical review of their relationship to the relevant parts of the intellectual heritage of humanity, with logical tests of the validity of such relationship.

- Paradigmatic validity; this may be applicable, but not necessary, since interdisciplinary and philosophical theories are above paradigms.

- A design of critical experiments if such are possible, aimed at checking whether the tested theory provides essential new insights.

- A design of descriptive experiments if the theory has descriptive character, and would be aimed at checking whether the tested theory describes reality accurately.

- Examples of applications, where the tested theory allows for prescriptive conclusions, and is aimed at checking whether applications confirm expected impacts of prescribed actions.

8.3.1 General spiral of knowledge creation

A general spiral of evolutionary knowledge creation (the Observation, Enlightenment, Application, Modification, or OEAM spiral) might thus proceed along the following pathway. See also Figure 8.4.

1. *Observation*: We observe reality (either in nature or in society) and its changes, in order to compare our observations with knowledge of human heritage (follows postmodern subjectivism).

2. *Enlightenment*: Then our intuitive and emotive knowledge helps us generate new hypotheses (follows paradigmatism).

3. *Application*: We create new tools by applying them to existing reality (follows falsificationism).

4. *Modification*: Usually with the goal of achieving some changes, or modifications to reality, we observe them again.

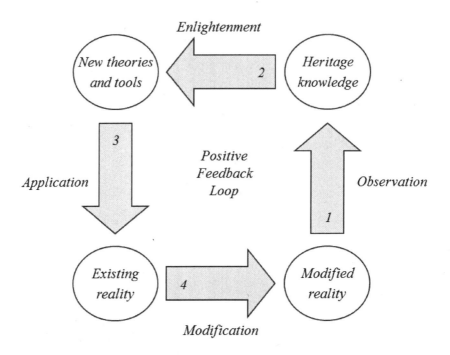

Figure 8.4: A general OEAM spiral (Wierzbicki and Nakamori, 2008).

8.3.2   Episteme: exercise

Exercise 8.1

1. Give examples of the OEAM spiral.

2. Or, consider a combination of three epistemological positions:

   - paradigmatism
   - falsificationism
   - postmodern subjectivism

8.4 Theory of Knowledge Construction Systems

We have discussed the *Informed Systems Approach* in Chapter 3, which consists of three principles:

1. The principle of cultural sovereignty

2. The principle of informed responsibility

3. The principle of systemic integration

As a concrete methodology of the *Informed Systems Approach*, a knowledge integration system called the *i*-System was introduced in Chapter 3. The necessary agencies in collecting knowledge were considered in Chapter 7. In this chapter we consider a new episteme, called *evolutionary constructive objectivism*, for justifying collected and integrated knowledge. It is now appropriate to introduce the theory of knowledge construction systems (Nakamori, Wierzbicki, and Zhu, 2011). See Figure 8.5.

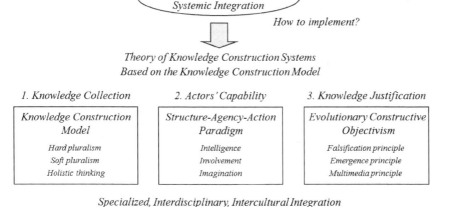

Figure 8.5: Theory of knowledge construction systems (summary).

8.4.1   Fundamental parts of the theory

The theory consists of three fundamental parts (Nakamori, 2012):

- *The knowledge integration system*: A basic system to collect and synthesize various knowledge, called the *i*-System, which itself is a systems methodology (Nakamori, 2000, 2003).

- *The structure-agency-action paradigm*: A sociological interpretation of the *i*-System to emphasize the necessary abilities of actors when collecting and synthesizing knowledge (Nakamori and Zhu, 2004).

- *Evolutionary-constructive objectivism*: A new episteme to create knowledge and justify collected and synthesized knowledge (Wierzbicki and Nakamori, 2007a, 2007b, 2008).

8.4.2   Evolutionary constructive objectivism

The three basic principles that will be necessary for the change to the new episteme of the knowledge civilization era are:

- *Evolutionary Falsification Principle*: Hypotheses, theories, models and tools evolve, and the measure of their evolutionary fitness is the number of either attempted falsification tests that they have successfully passed, or of *critical discussion tests* leading to an inter-subjective agreement about their validity.

- *Emergence Principle*: New properties of a system emerge with increased levels of complexity, and these properties are qualitatively different from, and irreducible to, the properties of the system's parts. We should not hesitate to make a new concept from insight.

- *Multimedia Principle*: Words are just an approximate code to describe a much more complex reality. Visual and nonverbal information in general is much more powerful and relates to intuitive knowledge and reasoning. We should stimulate maximum creativity using multimedia.

Epistemological position

- Based on hypotheses about reality, we create diverse models of the world. All such hypotheses turn out to be only approximations, but we test their validity by using the *falsification principle*.

- Since we perceive reality as more and more complex, and devise concepts on higher levels of complexity according to the *emergence principle*, we shall probably always work with approximate hypotheses.

- According to the *multimedia principle*, language is a simplified code used to describe a much more complex reality, while human senses enable people to perceive the more complex aspects of reality. This more comprehensive perception of reality is the basis of human intuition.

Knowledge justification   Why knowledge justification rather than verification? Because time is necessary for verification of new knowledge; it has to be justified before putting it into practice. Table 8.1 has been prepared to evaluate activities in knowledge creation. Reflecting such an evaluation, we can make a new improved activity plan.

Table 8.1: Evaluation table for knowledge synthesis activities.

i-System	*Grade*	*Agency*	*Grade*	*Justification*	*Grade*
Hard pluralism	*Average*	*Intelligence*	*Good*	*Falsification*	*Good*
Soft pluralism	*Poor*	*Involvement*	*Excellent*	*Emergence*	*Poor*
Holistic thinking	*Good*	*Imagination*	*Average*	*Multimedia*	*Fair*

Poor < Fair < Average < Good < Excellent

8.5   Application of Intercultural Integration

Let us return to the problem of regional revitalization discussed in Section 1.3 and also in Section 3.4. We investigated a revitalization plan in a small city in Japan. The city had several plans, among which we focused on its biomass town plan, which included the utilization of garbage, green waste, used cooking oil, sewage sludge, and food processing waste. This city developed a promotional program that has not, however, led to a widespread uptake at this time. What are the problems or obstacles? We tried to explore these with the *i*-System.

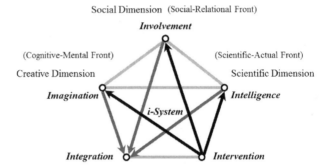

Knowledge collection and synthesis

Intervention: We decided to explore the reasons for slow progress in establishing a biomass town, focusing exclusively on the profitable use of garbage.

Intelligence: We collected scientific knowledge from researchers and government officials, who were key people in promoting this business. We found that there was no quantitative analysis to estimate effects of this activity on the regional environment and economy, and that there was little research about the quality of compost and ingredients.

Involvement: Municipal officials introduced us to a trader of recycled materials, civic groups, and farmers who had become the core group promoting the biomass town plan. We interviewed them to learn about their current involvement, dissatisfactions or expectations in promoting the plan. They had already established a scheme of food circulation and resource recycling as shown in Figure 8.6.

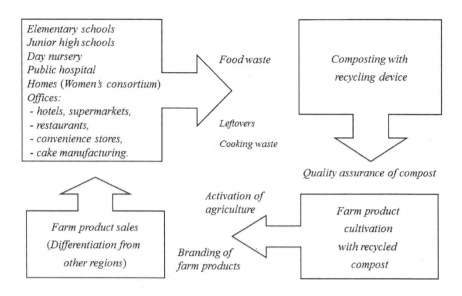

Figure 8.6: A scheme of food circulation and resource recycling.

Imagination: By sending a questionnaire to 2000 citizens, we collected their opinions and hopes, (with a 42% response rate). Using a structuring method (affinity diagram) for opinions, we summarized these as shown in Section 1.3.

Integration: We assembled knowledge from the three dimensions (or fronts) and summarized the problems in promoting the biomass town plan.

Self-evaluation　Here let us summarize the self-evaluation of our activity in this investigation.

- In *Intelligence*, we could use systems analysis (optimization or simulation, scenario analysis, etc.), or research management (or technology management) to obtain more reliable information. These were left for future investigation. We cannot say that we used our ability of *Intelligence* to the greatest extent possible; but within a restricted situation we used our *Intelligence* considerably. We accepted the opinions of the researchers and the officials without attempting to put forward counter-evidence; which means that we did not follow the falsification principle.

- In *Involvement*, we interviewed almost all the people who had already taken part in the food circulation and resource recycling system. But we did not try any other systems methodology, in order to find an emergence of opinions among relevant people. This means that we did not fully follow the soft pluralism and emergence principle.

- In *Imagination*, we ran a questionnaire survey for 2000 citizens. We used not only prepared questions but also asked them their opinions. The latter included opinions based on their holistic thinking on environmental problems, while we ourselves thought about the problem holistically only when summarizing opinions of relevant people. However, we did not explain to the citizens the status of the current plan using multimedia.

Our self-evaluation of this investigation can be summarized as in Table 8.2, which is important because it can be reflected in the next project. For instance, we can proceed to the next research plan using the *i*-System to build a business model as shown in Figure 8.7.

Table 8.2: A grade sheet of self-evaluation.

i-System	*Grade*	*Agency*	*Grade*	*Justification*	*Grade*
Hard pluralism	*Fair*	*Intelligence*	*Good*	*Falsification*	*Poor*
Soft pluralism	*Poor*	*Involvement*	*Excellent*	*Emergence*	*Poor*
Holistic thinking	*Average*	*Imagination*	*Good*	*Multimedia*	*Poor*

Poor < Fair < Average < Good < Excellent

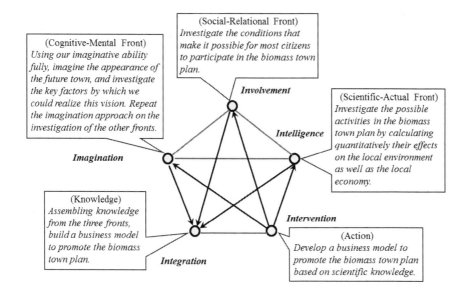

Figure 8.7: A research plan to build an environmental business model.

8.6 Final Exercise

Exercise 8.2 In your master's or doctoral research (or in your daily life), how do you discover or create knowledge, and how do you justify it?

1. What kind of knowledge creation model will you use?

2. What kinds of knowledge will you collect?

3. From where and how will you collect such knowledge?

4. How will you synthesize collected knowledge?

5. How will you justify (or verify) your created knowledge?

Amazing!
You have reached
the final exercise.

Intercultural integration of knowledge is a
decisive concept of the new systems science
in the knowledge-based society.

| Knowledge creation | → | *There is something noble in creating new knowledge.* |

| Knowledge synthesis | → | *There is strong pressure to be perfect.* |

| Knowledge construction | → | *There is reassurance in being able to repeat something* |

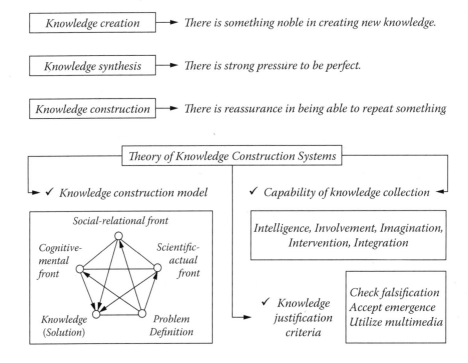

References

Ackerman MS, Pipek V, Wulf V eds. (2003) *Sharing Expertise: Beyond Knowledge Management*, MIT Press, Cambridge.

Ackoff RL (1974) *Redesigning the Future*, John Wiley & Sons, New York.

Ackoff RL (1978) *The Art of Problem Solving*, John Wiley & Sons, New York.

Ackoff RL (1981) *Creating the Corporate Future*, John Wiley & Sons, New York.

Archer MS (1995) *Realist Social Theory: The Morphogenetic Approach*, University of Cambridge Press, Cambridge.

Arthur WB (1988) Self-reinforcing mechanism in economics, in Anderson P et al. eds., *The Economy as an Evolving Complex System*, Addison-Wesley, Reading.

Ashby WR (1956) *An Introduction to Cybernetics*, Chapman & Hall, London.

Ashby WR (1958) Requisite variety and its implications for the control of complex systems, *Cybernetica* 1(2), 83–99.

Beer S (1972) *Brain of the Firm*, Allen Lane, London (Second edition, 1981, John Wiley & Sons, Chichester).

Beer S (1979) *The Heart of Enterprise*, John Wiley & Sons, Chichester.

Beer S (1985) *Diagnosing the System for Organizations*, John Wiley & Sons, Chichester.

Beer S (1994) *Beyond Dispute: The Invention of Team Syntegrity*, John Wiley & Sons, Chichester.

Beishon J, Peters G (1972) *Systems Behaviour*, Harper & Row, London.

Bergson H (1903) *Introduction to Metaphysics*, English translation first published in 1911, reprinted by Hackett Publishing Company, 1999, Indianapolis.

Bertalanffy LV (1956) General system theory, *General Systems* 1, 1–10.

Bertalanffy LV (1968) *General Systems Theory: Foundations, Development, Application*, Penguin, Harmondsworth.

Blanchard BS, Fabrycky WJ (2010) *Systems Engineering and Analysis* (Fifth edition), Prentice Hall, Upper Saddle River, New Jersey.

Britz GC et al. (2000) *Improving Performance through Statistical Thinking*, ASQ Quality Press, London.

Buckley W (1965) *Modern Systems Research for the Behavioral Scientist*, Aldine, Chicago.

Checkland PB (1978) The origins and nature of "hard" systems thinking, *Journal of Applied Systems Analysis* **5**(2), 99–110.

Checkland PB (1981) *Systems Thinking, Systems Practice*, John Wiley & Sons, New York.

Checkland PB, Scholes J (1990) *Soft Systems Methodology in Action*, John Wiley & Sons, Chichester.

Cheng CY (1997) The origins of Chinese philosophy, in Carr B, Mahalingam I eds., *Comparison Encyclopedias of Asian Philosophy*, Routledge, London.

Child J (1997) Strategic choice in the analysis of action, structure, organisations and environment: Retrospect and prospect, *Organisation Studies* **18**(1), 43–76.

Choquet G (1954) Theory of capacities, *Annales de L'institut Fourier* **5**, 131–295.

Comte A (1848) *Discours sur l'ensemble du positivisme*, Translation by Bridges JH in 1865, *A General View of Positivism*, Trubner and Co., London (reissued by Cambridge University Press, 2009).

David PA (1985) Clio and the economics of QWERTY, *American Economic Review* **75**(2), 332–337.

Dierkes M, Child J, Antal AB, Nonaka I (2003) *Handbook of Organizational Learning and Knowledge*, Oxford University Press, Oxford.

Einstein A (1916) *Relativity: The Special and General Theory* (Translation in 1920), Henry Holt and Company, New York.

Emery FE (1969) *Systems Thinking, Volume One*, Penguin, Harmondsworth.

Emery FE (1981) *Systems Thinking, Volume Two*, Penguin, Harmondsworth.

Feigenbaum EA, McCorduck P (1983) *The Fifth Generation*, Addison-Wesley, Reading.

Flood RL, Carson ER (1993) *Dealing with Complexity: An Introduction to the Theory and Application of Systems Science* (Second edition), Plenum Press, New York.

Flood RL, Jackson MC (1991) *Critical Problem Solving: Total Systems Intervention*, John-Wily & Sons, Chichester.

Foucault M (1972) *The Order of Things: An Archeology of Human Sciences*, Routledge, New York.

Fu CW (1997) Daoism in Chinese philosophy, in Carr B, Mahalingam I eds., *Comparison Encyclopedias of Asian Philosophy*, Routledge, London.

Gasson S (2004) The management of distributed organizational knowledge. In *Proceedings of the 37th Hawaii International Conference on Systems Sciences*, Track 8, January 5–8, Hawaii.

Garud R, Rappa MA (1994) A socio-cognitive model of technology evolution: The case of cochlear implants, *Organisation Science* **5**(3), 344–362.

Giddens A (1979) *Central Problems in Social Theory: Action, Structure and Contradiction in Social Analysis*, Macmillan, London.

Grabisch M (1996) The application of fuzzy integrals in multicriteria decision making, *European Journal of Operational Research* **89**(3), 445–456.

Grabisch M (2003) Modelling data by the Choquet integral, in Torra V ed., *Information Fusion in Data Mining* (*Studies in Fuzziness and Soft Computing* **123**), 135–148.

Gu JF (2012) Knowledge synthesis, in Nakamori Y ed. *Knowledge Science: Modeling the Knowledge Creation Process*, Chapter 5, 81–102, CRC Press, London.

Gu JF, Tang XJ (2005) Meta-synthesis approach to complex system modeling, *European Journal of Operational Research* **166**(3), 597–614.

Gu JF, Zhu ZC (2000) Knowing Wuli, sensing Shili, caring for Renli: Methodology of the WSR approach, *Systemic Practice and Action Research* **13**(1), 11–20.

Habermas J (1972) *Knowledge and Human Interests*, Shapiro JJ trans., Heinemann, London.

Habermas J (1987) *Lectures on the Philosophical Discourse of Modernity*, MIT Press, Cambridge, Massachusetts.

Hammond D (2003) *The Science of Synthesis: Exploring the Social Implications of General Systems Theory*, University Press of Colorado, Colorado.

Hanson BG (2002) *General Systems Theory: Beginning with Wholes*, Taylor & Francis, Washington, D.C.

Heisenberg W (1927) Ueber den anschaulichen Inhalt der quantentheoretischen Kinematik and Mechanik, *Zeitschrift für Physik* **43**, 172–198. English translation in 1983 by Wheeler JA, Zurek WH, 62–84.

Hirschfeld HO (1935) A connection between correlation and contingency, in *Mathematical Proceedings of the Cambridge Philosophical Society* **31**, 520–524.

Huang Z (1998) Extensions to the k-means algorithm for clustering large data sets with categorical values, *Data Mining and Knowledge Discovery* **2**, 283–304.

Huynh VN, Yan HB, Nakamori Y (2010) A target-based decision making approach to consumer-oriented evaluation model for Japanese traditional craft, *IEEE Transactions on Engineering Management* **57**(4), 575–588.

Jackson MC (1991) *Systems Methodology for the Management Science*, Plenum Publishers, New York.

Jackson MC (2000) *Systems Approaches to Management*, Kluwer/Plenum, New York.

Jackson MC (2003) *Systems Thinking: Creative Holism for Managers*, John Wiley & Sons, Chichester.

Jackson MC, Keys P (1984) Towards a system of systems methodologies, *Journal of the Operational Research Society* **35**(6), 473–486.

Johnston J (2008) *The Allure of Machinic Life: Cybernetics, Artificial Life, and the New AI*, MIT Press, Cambridge.

Kant I (1951) *Critique of Judgment*, Translated by Bernard JH, Hafner Publishing, New York (original publication date 1892).

Kawakita J (1967) *Idea Creation (Way of Thinking): For the Development of Creativity*, Chukoshinsho, Tokyo (in Japanese).

Kendall KE, Kendall JE (2010) *Systems Analysis and Design* (Eighth edition), Prentice Hall, New Jersey.

Kikuchi T, Rong LL, Wang ZT, Wierzbicki AP, Nakamori Y (2007) Evaluation of research capabilities and environments in academia based on a knowledge creation model, *International Journal of Knowledge and Systems Sciences*, **4**(1), 14–24.

Klir G (1991) *Facets of Systems Science*, Plenum, New York.

Kostoff RN, Boylan R, Simons GR (2004) Disruptive technology roadmaps, *Technological Forecasting and Social Change* **71**, 141–159.

Krogh GV, Ichijo K, Nonaka I (2000) *Enabling Knowledge Creation: How to Unlock the Mystery of Tacit Knowledge and Release the Power of Innovation*, Oxford University Press, Oxford.

Król Z (2007) The emergence of new concepts in science, in Wierzbicki AP, Nakamori Y eds., *Creative Environments: Issues of Creativity Support for the Knowledge Civilization Age*, Chapter 17, 417–444, Springer, Berlin.

Kuhn TS (1962) *The Structure of Scientific Revolutions*, Chicago University Press, Chicago.

Kuhn TS (1970) *The Structure of Scientific Revolutions* (Second edition), University of Chicago Press, Chicago.

Kuhn TS (2000) *The Road after "Structure,"* University of Chicago Press, Chicago.

Lendaris GG (1980) Structural modeling: A tutorial guide, *IEEE Transactions on Systems, Man, and Cybernetics* **10**(12), 807–840.

Liebowitz J ed. (2012) *Knowledge Management Handbook: Collaboration and Social Networking* (Second edition), CRC Press, Boca Raton.

Lindlof TR, Taylor BC (2002) *Qualitative Communication Research Methods* (Second edition), Sage Publications, Thousand Oaks.

Linstone HA, Turoff M eds. (1975), *The Delphi Method: Techniques and Applications*, Addison-Wesley, Reading.

Lyotard JF (1984) *The Postmodern Condition: A Report on Knowledge*, Manchester University Press, Manchester.

Ma T, Liu S, Nakamori Y (2005) Roadmapping as a way of knowledge management for supporting scientific research in academia, *Systems Research and Behavioral Science* **22**, 1–13.

MacQueen JB (1967) Some methods for classification and analysis of multivariate observations, in *Proceedings of the 5th Symposium on Mathematical Statistics and Probability* **1**, 281–297, Berkeley.

Maturana HR, Varela FJ (1998) *The Tree of Knowledge: The Biological Roots of Human Understanding*, Shambhala Publications, Boston.

Medina E (2011) *Cybernetic Revolutionaries: Technology and Politics in Allendes Chile*, MIT Press, Cambridge.

Midgley G (2000) *Systems Intervention: Philosophy, Methodology and Practice*, Kluwer/Plenum, New York.

Midgley G ed. (2003) *Systems Thinking*, Volumes 1, 2, 3, 4, Sage Publications, London.

Midgley G (2004) Systems thinking for the 21st century. *International Journal of Knowledge and Systems Sciences* **1**(1), 63–69.

Mulej M (2007) Systems theory: A world view and/or a methodology aimed at requisite holism/realism of human's thinking, decisions and action, *Systems Research and Behavioral Science* **24**(3), 347–357.

Nahapiet JE, Ghoshal S (1998) Social capital, intellectual capital, and the organizational advantage, *Academy of Management Review* **23**(2), 242–266.

Nakamori Y (2000) Knowledge management system toward sustainable society, In *Proceedings of the 1st International Symposium on Knowledge and System Sciences* 57–64, September 25–27, Ishikawa, Japan.

Nakamori Y (2003) Systems methodology and mathematical models for knowledge management, *Journal of Systems Science and Systems Engineering* **12** (1), 49–72.

Nakamori Y (2011) Kansei information transfer technology, in *Proceedings of International Symposium on Integrated Uncertainty in Knowledge Modeling and Decision Making*, 209–218, October 28–30, 2011, Hangzhou, China.

Nakamori Y ed. (2012) *Knowledge Science: Modeling the Knowledge Creation Process*, CRC Press, London.

Nakamori Y, Ryoke M (2006) Treating fuzziness in subjective evaluation data, *Information Sciences* **176**, 3610–3644.

Nakamori Y, Sawaragi Y (1990) Shinayakana Systems Approach in environmental management, in *Proceedings of the 11th World Congress of International Federation of Automatic Control* **5**, 511–516, August 13–17, Tallinn, USSR.

Nakamori Y, Sawaragi Y (1992) Shinayakana Systems Approach to modeling and decision support, in *Proceedings of the 10th International Conference on Multiple Criteria Decision Making*, Vol. I, 77–86, , July 19–24, Taipei, Taiwan.

Nakamori Y, Wierzbicki AP, Zhu ZC (2011) A theory of knowledge construction systems, *Systems Research and Behavioral Science* **28**, 15–39.

Nakamori Y, Zhu ZC (2004) Exploring a sociologist understanding for the *i*-System, *International Journal of Knowledge and Systems Sciences* **1**(1), 1–8.

Nagamachi M, ed. (2011) *Kansei/Affective Engineering*, CRC Press, Boca Raton.

Nagamachi M, Lokman AM (2011) *Innovation of Kansei Engineering*, CRC Press, Boca Raton.

Nisbett RE (2003) *The Geography of Thought*, Nicholas Brealey Publishing, London.

Nonaka I ed. (2005) *Knowledge Management: Critical Perspectives on Business and Management*, Volumes I, II, III, Routledge, London.

Nonaka I, Takeuchi H (1995) *The Knowledge-Creating Company: How Japanese Companies Create the Dynamics of Innovation*, Oxford University Press, Oxford.

Nonaka I, Zhu ZC (2012) *Pragmatic Strategy: Eastern Wisdom, Global Success*, Cambridge University Press, Cambridge.

North DC (1990) *Institutions, Institutional Change and Economic Performance*, Cambridge University Press, New York.

Ohn Mar San, Huynh VN, Nakamori Y (2003) A clustering algorithm for mixed numeric and categorical data, *Journal of Systems Science and Complexity* **16**(4), 562–571.

Orlikowski WJ (2002) Knowing in practice: Enacting a collective capability in distributed organizing, *Organization Science* **13**(3), 249–273.

Osborn AF (1953) *Applied Imagination: Principles and Procedures of Creative Problem Solving*, Charles Scribner's Sons, New York.

Osgood CE, Suci GJ, Tannenbaum PH (1957) *The Measurement of Meaning*, University of Illinois Press, Urbana.

Parsons T (1950) *The Social System*, Free Press, New York.

Phaal R, Farrukh C, Probert D (2001) *T-plan: Fast Start to Technology Roadmapping: Planning Your Route to Success*, University of Cambridge, Institute for Manufacturing, Cambridge.

Pickering A (2010) *The Cybernetic Brain: Sketches of Another Future*, University of Chicago Press, Chicago.

Poincaré H (1913) *The Foundations of Science* (English translation in 1946), Science Press, Lancaster.

Polanyi M (1958) *Personal Knowledge: Towards a Post-Critical Philosophy*, Routledge & Kegan Paul, London.

Polanyi M (1966) *The Tacit Dimension*, Routledge & Kegan Paul, London.

Popper KR (1934) *Logik der Forschung*, Julius Springer Verlag, Vienna.

Rescher N (1998) *Predicting the Future: An Introduction to the Theory of Forecasting*, State University of New York Press, New York.

Ryle G (1949) *The Concept of Mind* (Current edition 1984), University of Chicago Press, Chicago.

Ryoke M, Yamashita Y, Hori K, Nakamori Y (2007) Knowledge discovery of interview survey on fresh food management, *International Journal of Knowledge and Systems Sciences* **4**(1), 31–34.

Saaty TL (1980) *Analytical Hierarchy Process*, McGraw-Hill, New York.

Saaty TL (1990) How to make a decision: The analytic hierarchy process, *European Journal of Operational Research* **48**, 9–26.

Schon DA (1983) *The Reflective Practitioner: How Professionals Think in Action*, Basic Books, New York.

Schumpeter JA (1934) *The Theory of Economic Development: An Inquiry into Profits, Capital, Credit, Interest and the Business Cycle*, Harvard University Press, Cambridge.

Scott WR (1995) *Institutions and Organisations*, Sage, Thousand Oaks.

Selye H (1964) *From Dream to Discovery*, McGraw Hill, New York.

Sewell WH Jr (1992) A theory of structure: Duality, agency, and transformation, *American Journal of Sociology* **98**(1), 1–29.

Skyttner L (2001) *General Systems Theory: Ideas and Applications*, World Scientific Publishing, Singapore.

Skyttner L (2006) *General Systems Theory: Problems, Perspectives, Practice*, World Scientific Publishing, Singapore.

Spender JC (1996) Organizational knowledge, learning and memory: Three concepts in search of a theory, *Journal of Organizational Change Management* **9**(1), 63–78.

Taha HA (2010) *Operations Research: An Introduction* (Ninth edition), Prentice Hall, New Jersey.

Taket AR, White LA (2000) *Partnership and Participation: Decision-Making in the Multiagency Setting*, John Wiley & Sons, Chicheter.

Takeuchi H, Nonaka I (2004) *Hitotsubashi on Knowledge Management*, John Wiley & Sons, Singapore.

Tian J, Nakamori Y, Wierzbicki AP (2009) Knowledge management and knowledge creation in academia: A study based on surveys in a Japanese research university, *Journal of Knowledge Management* **13**(2), 76–92.

Ulrich W (1983) *Critical Heuristics of Social Planning: A New Approach to Practical Philosophy*, Haupt, Bern.

Warfield JN (1974) Toward interpretation of complex structural models, *IEEE Transactions on Systems, Man, and Cybernetics* **4**(5), 405–417.

Warfield JN (1976) *Societal Systems: Planning, Policy and Complexity*, John Wiley & Sons, New York.

Wiener N (1948) *Cybernetics: Control and Communication in the Animal and the Machine*, MIT Press, Cambridge.

Wierzbicki AP, Nakamori Y (2006) *Creative Space: Models of Creative Processes for the Knowledge Civilization Age*, Springer, Berlin.

Wierzbicki AP, Nakamori Y (2007a) The episteme of knowledge civilization, *International Journal of Knowledge and Systems Sciences* **4**(3), 8–20.

Wierzbicki AP, Nakamori Y, eds. (2007b) *Creative Environments: Issues of Creative Support for the Knowledge Civilization Age*, Springer, Berlin.

Wierzbicki AP, Nakamori Y (2008) The importance of multimedia principle and emergence principle for the knowledge civilization age, *Journal of Systems Science and Systems Engineering* **17**(3), 297–318.

Wierzbicki AP, Zhu ZC, Nakamori Y (2006) A new role of systems science: Informed Systems Approach, in Wierzbicki AP, Nakamori Y eds., *Creative Space: Models of Creative Processes for the Knowledge Civilization Age*, Chapter 6, 161–215, Springer, Berlin.

Wikipedia (2013a) Affinity diagram, http://en.wikipedia.org/wiki/Affinity_diagram (March 24, 2013).

Wikipedia (2013b) Delphi method, http://en.wikipedia.org/wiki/Delphi_method (March 24, 2013).

Wikipedia (2013c) Homeostasis, http://en.wikipedia.org/wiki/Homeostasis (March 24, 2013).

Wikipedia (2013d) Recommender system, http://en.wikipedia.org/wiki/Recommender_system (March 24, 2013).

Wilson TD (2002) The nonsense of "knowledge management," *Information Research* **8**(1): http://informationr.net/ir/8-1/paper144.html.

Yager RR (1988) On ordered weighted averaging aggregation operators in multicriteria decision making, *IEEE Transactions on Systems, Man, and Cybernetics* **18**(1), 183–190.

Yager RR, Kacprzyk J (1997) *The Ordered Weighted Averaging Operators: Theory and Applications*, Kluwer, Norwell.

Yamashita Y, Nakamori Y (2007) Knowledge integration methodology for designing a knowledge base of technology development in traditional craft industry, in *Proceedings of the 2007 IEEE International Conference on Systems, Man, and Cybernetics*, 332–337, October 7–10, Montreal, Canada.

Zadeh LA (1965) Fuzzy sets, *Information and Control* **8**(3), 338–353.

Zadeh LA (1983) A computational approach to fuzzy quantifiers in natural languages, *Computers and Mathematics with Applications* **9**(1), 149–184.

Zhu ZC (1998) Confucianism in action: Recent developments in Oriental systems methodology, *Systems Research and Behavioural Science*, **15**, 111–130.

Zhu ZC (2000) Dealing with a differentiated whole: The philosophy of the WSR approach, *Systemic Practice and Action Research* **13**(1), 21–57.

Index